Viroinformatics and Computational Virology in Public Health

Hyeon Seok Son • Myeongji Cho

Viroinformatics and Computational Virology in Public Health

Springer

Hyeon Seok Son
Graduate School of Public Health
Seoul National University
Seoul, Korea (Republic of)

Myeongji Cho
Division of Healthcare
and Artificial Intelligence
National Institute of Health, Korea
Disease Control and Prevention Agency
Cheongju-si, Korea (Republic of)

ISBN 978-981-96-5756-8 ISBN 978-981-96-5757-5 (eBook)
https://doi.org/10.1007/978-981-96-5757-5

This Springer imprint is published by the registered company Springer Nature Singapore Pte Ltd.
The registered company address is: 152 Beach Road, #21-01/04 Gateway East, Singapore 189721, Singapore

Preface

The foundation of viroinformatics lies in advancements in computer hardware and software technologies. Computational techniques have been applied in biology for quite some time. However, the turning point that ingrained them in the public consciousness and led biologists to actively recognize their necessity was the establishment of the Human Genome Project. Discussions about this project began in the United States and spread worldwide, leading scientists in various countries to seek government funding, often successfully, to conduct new genomic research. The private company Celera Genomics challengingly joined the sequencing race, accelerating the announcement of genomic results 2 years ahead of schedule. Celera's strategy involved using high-performance computers to rapidly decode sequences. Consequently, in June 2000, the successful completion of a draft sequence of the human genome was announced. Notably, the field of computer science, previously considered distant from life sciences, was recognized as an essential component of this achievement. Thus, the analysis and synthesis of biological phenomena through computers were destined to increase. The development of big data technology and artificial intelligence may lead life sciences in unexpected directions in the future. However, it can confidently be stated that information processing technology using computers will allow for the rapid extraction of useful information from biological data, and many biological phenomena will fall into predictable categories.

The three core elements of bioinformatics can be defined as storing, processing, and communication, with biological data as the objects. Processing data to produce information is widely considered a fundamental goal. Various tools and protocols are employed to achieve this goal, and the foundational elements of each can be considered its technical components. Examples include programming languages, database languages, and the internet. The reasons for studying bioinformatics may vary, but the processes involved fall within these fields. Technically, the first logical step is to build a database to store data properly. The second is to process the stored data for specific purposes, which requires programming skills. Various programming languages are available, and while any can be chosen, commonly used languages like Python or Java are essential because of their accessibility and shareability with other researchers. Additionally, building databases for the produced or refined data presents a technical challenge that students might find interesting. However, in most cases, using websites built by other researchers or data providers is a more natural approach.

We have now reached a stage where computational techniques and bioinformatics can actively assist the field of virology. Regardless of whether viruses are viewed as living entities, the contributions of biology and bioinformatics to the field will be significant. Viruses persistently trouble humanity. The 1918 Spanish flu and the recent COVID-19 pandemic are prime examples of viral infectious diseases. The COVID-19 pandemic came as a surprise to a society that believed that medical and scientific advancements would soon conquer infectious diseases, causing immense suffering. Experience has shown that the threat of viral infections is not only a medical and public health issue but also a social crisis.

The need for international cooperation transcending regional boundaries is evident in today's highly interconnected world. We require a broader understanding of global challenges and response strategies that go beyond conventional perceptions alongside a positive recognition of interdisciplinary holistic approaches. I hope viroinformatics can significantly contribute to such efforts.

How to Approach This Book

This book aims to apply computational science methodologies, enhanced by advancements in computer science, to virology, an essential health sciences field. Biology and biotechnology have undeniably evolved alongside data science. Viroinformatics can be framed similarly to bioinformatics. Mastering (molecular) biology, bioinformatics, and viroinformatics is challenging, and expecting a few books to enable this achievement may be overly ambitious. However, this book was conceived from the understanding that every endeavor needs a starting point. Researching viruses requires a blend of natural science and medical perspectives, and health students must also embrace a comprehensive social viewpoint, as the threats posed by viruses are both real and potentially devastating.

The flow of this book does not deviate significantly from the conventional approach of learning, progressing from basic knowledge and skills to applications. However, this book is not intended to provide complete knowledge and skills. Such a goal could not be achieved rapidly and, if attempted, would necessitate an encyclopedic level of completeness, leaving the book eternally unfinished. Therefore, I encourage readers to recognize the practical applications of viroinformatics through this book and use the keywords introduced in each chapter as guides for further study. Acquiring practical knowledge and skills requires more than simple learning; it involves embodiment and training. To that end, I suggest reading as many texts as possible based on the clues gathered from this book. Additionally, I encourage gaining as much practical experience as possible. Excellent resources for this purpose can easily be found on the internet or through other sources.

This book is divided into three sections. Roughly speaking, these three parts together form the framework of the field defined as viroinformatics. The first section, "Prerequisites for Viroinformatics," serves as an introduction, covering basic concepts of virology, molecular biology related to viruses, and bioinformatics. Detailed techniques and applications are covered in the subsequent sections. Readers familiar with viruses are encouraged to skip this section to save time. The second section, entitled "Essential Methods for Viroinformatics," mainly introduces methodologies in the bioinformatics domain. The relevant principles, along with examples, are presented. The third section, "Applications and Case Studies," describes cases where the methodologies introduced in the previous section are applied individually or in combination. It should be noted that the research area focuses on health, differing somewhat from purely scientific research. For instance, in a case study using phylogenetic analysis, other elements, such as alignment methods, are inevitably involved, but detailed explanations of these are omitted. Focusing on how the algorithms or methodologies used address the presented problems will achieve the intended purpose.

I hope readers can elevate their understanding using the concepts, methodologies, and keywords introduced in this book. No research or application places importance solely on adhering to standardized or illustrated examples. All progress lies in appropriate and creative applications.

Seoul, Korea (Republic of) Hyeon Seok Son
Cheongju-si, Korea (Republic of) Myeongji Cho
December 2024

Acknowledgments

We would like to express our deepest gratitude to everyone who supported us throughout the creation of this book. First and foremost, we thank each other for the opportunity to collaborate on this project. Working together has been an inspiring and rewarding experience. We both greatly admire each other's expertise, passion, and dedication to our shared work and interests. This collaboration has truly been a highlight of our professional lives.

Professor Hyeon Son would particularly like to thank Dr. Myeongji Cho for her steadfast support, insightful contributions, and dedication to the success of this book. He is confident that her expertise, intellectual curiosity, and collaborative spirit have made this project both enjoyable and meaningful. He is deeply grateful for her commitment to this work and her invaluable partnership throughout the writing process.

Professor Son would also like to express his sincere gratitude to his past mentors, Professors John L. Martin, R. J. Sheppard, and Mark S. P. Sansom. Their guidance, wisdom, and encouragement played a pivotal role in shaping his academic journey and approach to research. Their support continues to inspire and influence his work today.

Dr. Cho would like to extend her warm appreciation to Professor Son for the opportunity to co-author this important book. Dr. Cho has greatly valued his profound insight, mentorship, and sense of humor, which have made the writing process both enjoyable and enriching. His guidance, encouragement, and willingness to share his expertise have made this collaboration a truly rewarding experience.

We are both extremely grateful to our families for their constant support and encouragement. Professor Son would like to dedicate this book to the memory of his late parents, Sung I. Son and Keun N. Park, whose love, wisdom, and sacrifices have shaped the person he is today. Their enduring influence continues to inspire and guide him in all that he does.

Dr. Cho would like to thank J. A. Jo, Y. S. Shin, Sehoon, and Edam for their steadfast support, which made this journey possible.

Finally, we would like to thank all the colleagues, experts, and mentors who provided invaluable feedback and guidance along the way, as well as Springer Nature and its team for their role in bringing this work to life.

Contents

Part II Essential Methods for Viroinformatics

Part I

Prerequisites for Viroinformatics

1 Virology in Brief

About 100 years ago, viruses were recognized as independent entities distinct from other pathogens. The term *virus* originally referred to a disease-causing toxin. The existence of non-bacterial pathogens was doubted. Viruses that could pass through filters designed to trap bacteria were subsequently discovered, the first being the tobacco mosaic virus.

Viruses can infect all known living organisms. Unlike viruses that infect plants and animals, bacteriophages infect bacteria and were important subjects of virus research in the mid-twentieth century. A notable outcome from bacteriophage studies was the discovery that nucleic acids are the genetic material of viruses (Hershey and Chase 1952).[1]

Viruses can be said to exist everywhere around us. They are found both inside and outside our bodies, even in places one would not normally imagine. The belief that salty seawater can eliminate viruses is a misunderstanding. Fortunately, not all viruses are harmful to humans. The ability of viruses to infect and reproduce in all living organisms is remarkable. Even bacteria are not safe from viruses. Bacteriophages, which resemble space landing vehicles, are viruses that infect bacteria. Of course, each virus prefers a certain host species. Despite this species specificity, the barriers between species are not impenetrable. For example, zoonotic viruses are transmitted from animals to humans, and zoonotic potential is determined by the characteristics of the host and the virus.[2]

Retroviruses are remarkable viruses. Like other animals, human DNA contains retroviral DNA, and although this DNA does not have a marked negative effect on the host, it is passed down through generations. Although viruses may be harmful or at least not beneficial, they have contributed to biological research. Studies on viruses have led to the discovery of the functions of introns, as well as oncogenes and tumor-suppressor genes. Additionally, researchers are investigating the use of viral entry mechanisms to insert desired genes into cells deliberately using viral genomes, a process known as gene therapy.

Viruses can be defined as living entities and chemical substances, i.e., inanimate objects. While they cannot engage in life activities, including self-replication, viruses may be considered living organisms because they reproduce and transmit their genetic material to future generations by way of a host. Parasites do not lead completely independent lives. Therefore, it could

[1] In 1952, Hershey and Chase discovered that nucleic acids are the genetic material using T2 phage. Later, Fraenkel-Conrat and Singer confirmed that RNA is the genetic material using tobacco mosaic virus. This established that viruses could have DNA or RNA as their genetic material.

[2] Infections transmitted between animals and humans are referred to as zoonotic, and the resulting diseases are called zoonotic diseases.

H. S. Son, M. Cho, *Viroinformatics and Computational Virology in Public Health*,
https://doi.org/10.1007/978-981-96-5757-5_1

be argued that no living entity on Earth can survive and reproduce independently.

In the author's view, viruses are the smallest robots in the natural world observed by humans. They are fully automated machines with a unique purpose—limitless replication. Viruses prioritize the dissemination of their genetic material above all else, although since viruses cannot be seen as entities with a self, it would be incorrect to describe them as having such purposeful awareness. However, a reasonable personification that aids in understanding viruses could be useful. By exploiting host-cell functions, viruses continually reproduce. The death of cells is, to some extent, a result of explosive viral replication. The death of the host is not an ideal outcome for viruses, which implies that an evolutionary mechanism to delay cell death has developed. From this perspective, it is advantageous for viruses if all infections become endemic. Therefore, viruses rely on large host populations for their survival.

Because of the coronavirus disease 2019 (COVID-19) pandemic, much about viruses has become common knowledge, which when subjected to rigorous validation is a synonym for foundational knowledge. Those wishing to solve public health problems based on viroinformatics should arm themselves with knowledge that goes beyond common knowledge. Fortunately, there are many excellent texts available; however, it is not possible to summarise them all here. Therefore, I hope that you will seek out as many texts as possible and explore the history of humanity's struggle against viruses, because understanding factual history is foundational to all fields of study.

1.1 Classification of Viruses

Theoretically, there are countless ways to classify viruses. However, those that have most captured human interest are disease agents, and so early classifications were based on diseases. Classification can also be based on characteristics such as morphology, nucleic acid type, host organism, and phylogenetic relationships. For instance, viruses can be categorized by whether they contain DNA or RNA as their genetic material and whether they have an envelope. These traits are used for classification; for example, viruses with double-stranded DNA are referred to as dsDNA viruses, and those with an envelope among them are enveloped dsDNA viruses. Herpesviridae and Hepadnaviridae are enveloped dsDNA viruses, with the suffix -viridae indicating a family. Each classification method has utility depending on the context. The most widely used classification systems are outlined below.

1.1.1 The ICTV Classification

Taxonomy has been an important field of biology since the eighteenth century, when Carl Linnaeus defined the foundations of modern classification (Linné 1735). Taxonomy groups organisms based on unique identifiers formed from their intrinsic characteristics. The study of new organisms typically begins with comparative classification and naming in relation to other organisms (King et al. 2012). The principles, procedures, and nomenclature for the classification of animal species are handled by the International Commission on Zoological Nomenclature, and those for bacteria are determined by the International Committee on Systematic Bacteriology (Lapage et al. 1975; Ride 1999; Polaszek 2005).

Virus classification, which involves methods similar to ecological classification, is overseen by the International Committee on the Taxonomy of Viruses (ICTV). The International Committee on the Nomenclature of Viruses (ICNV) was established during the General Assembly of the International Association of Microbiological Societies (IAMS) in Moscow on July 22, 1966. A key decision at the founding meeting was that the ICNV would establish a classification and naming system for viruses distinct from that of bacteria. This led to a classification system for viruses that differs from existing rules in botany and zoology. In 1975, the ICNV became the ICTV, and in 1995, in response to the Internet age, a website was established to provide information (Adams et al. 2017). The ICTV consists of seven subcommittees[3] and has established over 100 international study groups (SGs) that focus on major virus families and genera. Each SG has a chair, who is appointed by the subcommittee chair. The responsibilities of the SG chair include organizing discussions among members on emerging classification issues, supervising the submission of proposals for new classifications, and overseeing the preparation and revision of the ICTV report.

The ICTV system is similar yet distinct from the general/traditional ecological species classification system. It maintains the order-family-subfamily-genus-species structure, with virus names followed by -virales (order), -viridae (family), -virinae (subfamily), -virus (genus), and species. According to the guidelines, virus names should not be italicized, and the words that make up a virus name are generally written in lowercase. However, in cases where the name includes a proper noun, such as a place name, or a combination of letters and numbers, a mixed case can be used (e.g., Zika virus, Human Immunodeficiency Virus Type 1, H1N1 influenza virus). Virus names may also be written as abbreviations that include both uppercase and lowercase letters, as well as numbers (e.g., SARS-CoV-2). In contrast to virus names, the naming of virus species is standardized by the ICTV. Virus species names must always be italicized, with the genus name starting with a capital letter (e.g., *Vesiculovirus indiana, Alphainfluenzavirus influenzae*). Following the binomial format, the second part of the species name may be written with a mixture of uppercase and lowercase letters or include numbers, and it may be freely chosen (e.g., *Plasmavirus L2*) (Zerbini et al. 2022). It is important to distinguish between "virus names" and "virus **species** names."

Sequence comparison is a key technique in virus classification, which is based on similarities and evolutionary relationships and has become increasingly important due to advances in computational technology. Sequence comparison considers various characteristics, including gene-expression mechanisms, host, pathogenicity, capsid structure and shape, and molecular composition. For example, all viruses of the order Picornavirales are non-enveloped viruses with positive-sense RNA and form icosahedral capsids. Picornavirales includes five families, one of which, the family Picornaviridae, has a single protein genome with a conserved genomic structure. Notably, the polyproteins of the Enterovirus genera have >42% amino acid identities, and Enterovirus C viruses share receptors that respond to a certain range of hosts. These established criteria are useful for classifying new viruses and identifying new lineages. Thus, viruses with similar characteristics are grouped to form hierarchical relationships (Le Gall et al. 2008; King et al. 2012; Carroll et al. 2019).

1.1.2 Baltimore Classification[4]

The Baltimore scheme, which is widely used alongside the ICTV classification, was proposed by David Baltimore in 1971. This classification considers the characteristics of viral genetic material, enabling the prediction of viral replication mechanisms. The two classifications are used complementarily. The Baltimore scheme, also known as the Baltimore Classification System, is conceptually straightforward. It is based on the fact that all viruses produce mRNA inside host cells. All known viruses require mRNA for replication, because it is essential for producing the necessary proteins and ultimately breaking out of the host cell. Therefore, classifying viruses based on the form of the source material and the manner in which mRNA is presented is a simple and rational approach. New emerging viruses can be handled according to this classification. In other words, as long as the focus on mRNA is unchanged, this classification will continue to be relevant.

[3]It consists of seven subcommittees that deal with animal DNA viruses and retroviruses, animal dsRNA and ssRNA(−) viruses, animal ssRNA(+) viruses, bacterial viruses, archaeal viruses, fungal and protist viruses, and plant viruses.

[4]David Baltimore (1938–) received the Nobel Prize in Physiology or Medicine in 1975 for his discovery of reverse transcriptase.

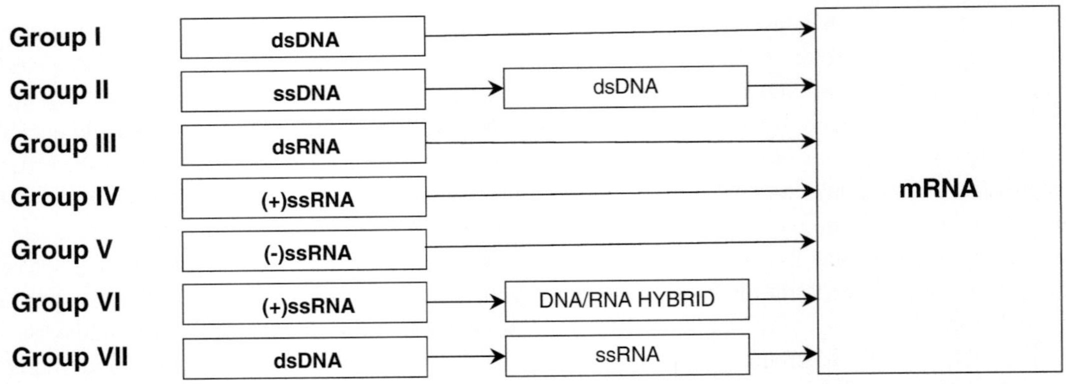

Fig. 1.1 The Baltimore scheme. The production of mRNA is a common characteristic of all viruses that invade host cells. All known viruses belong to one of seven categories in the Baltimore scheme
Group I (dsDNA): Viruses in this group have dsDNA. They use host enzymes to produce mRNA, although some species use viral enzymes. Examples include *Adenovirus* and *Papillomavirus*.
Group II (+ssDNA): These viruses have ssDNA, which is converted into dsDNA for use by host transcription enzymes to produce mRNA. Examples include *Circovirus* and *Parvovirus*.
Group III (dsRNA): These viruses have dsRNA and use their transcription enzymes to produce mRNA. An example is *Rotavirus*.
Group IV (+ssRNA): Viruses with positive-sense ssRNA genomes act directly as mRNA. Examples include *Poliovirus*, *Norovirus* and *Coronavirus*.
Group V (-ssRNA): These viruses have negative-sense ssRNA and use their transcription enzymes to produce mRNA. Examples include *Influenza virus* and *Rabies virus*.
Group VI (+ssRNA-RT): The RNA of these viruses is used for DNA production by viral reverse transcriptase. The DNA is integrated into the host genome and is used for mRNA production by host transcription enzymes. Examples include *Retroviruses* (such as *Human Immunodeficiency Virus* or *HIV*).
Group VII (dsDNA-RT): Viruses in this group have incomplete dsDNA, which is converted into complete dsDNA and ssRNA. Subsequently, dsDNA is synthesised, and mRNA is produced using host transcription enzymes. An example is *Hepatitis B virus*.

Figure 1.1 shows the Baltimore scheme, which categorizes viruses based on their characteristics up to the initial stage of mRNA production for replication within host cells. Therefore, assigning a virus to a classification group does not define all its characteristics, including replication and cell exit. Readers are encouraged to obtain information on the virus of interest.

1.2 Morphology and Structure

Viruses come in various shapes and sizes.[5] The functional unit of a virus is called a virion, which is considered a particle. Viruses contain genetic material—DNA or RNA—surrounded by a capsid to ensure its safe transport. The capsid is a protective protein shell that protects the virus from proteolytic enzymes. Viruses can be spherical, helical, or a combination of both, resembling lunar-lander-like structures (bacteriophages). These shapes are determined by the arrangement of the capsid; the capsid of spherical viruses is icosahedral.[6]

The coat that surrounds the capsid is a lipid membrane acquired from the plasma or intracellular membranes of the host cell during viral exit. This lipid bilayer membrane contains viral proteins, including glycoproteins. These membrane proteins play a crucial role in binding to host-cell receptors during infection (Uversky and Longhi 2011). Viruses with and without a lipid membrane are referred to as enveloped and naked viruses, respectively. The viral glycoprotein spikes in the coat fuse with the cell membrane

[5] Viruses typically range in size from 30–300 nm. However, the recent discovery of giant viruses has disproved the notion that all viruses are small. The Mimivirus, discovered in 1992, is the fourth largest virus after the Megavirus, Pandora virus, and Phytovirus. It was the first giant virus to be identified, with a capsid diameter of approximately 400 nm, and its host is amoebas.

[6] The only convex regular polyhedra that exist are the tetrahedron, cube, octahedron, dodecahedron, and icosahedron.

and so are essential for infection. Enveloped viruses, such as influenza virus, are pleomorphic, meaning that their virions vary in size and shape. Naked viruses are typically more stable than enveloped viruses. Although hand sanitizers can render enveloped viruses non-infectious, they do not affect naked viruses (e.g., norovirus). Therefore, proper handwashing is the best defense against infection by viruses.

1.3 Viral Proteins

Viral proteins are all proteins associated with a virus. Intrinsic viral proteins are proteins possessed by the virus prior to cell entry rather than those generated within the host cell or produced in response to the virus (e.g., antibodies). Therefore, intrinsic viral proteins are classified as viral structural proteins.

The viral capsid is composed of proteins. The capsid protects viral genetic material from environmental factors. It is composed of one or more proteins and is efficiently assembled, resulting in high packing efficiency. The proteins that make up the capsid form subunits. These come together to create intermediate units known as capsomers, a combination of *capsid* and *mere* (part in Greek), which aggregate to form the capsid. Capsids have a variety of shapes, but most are icosahedral or helical, with the former being more common than the latter.

The term capsid is derived from the Greek word *capsa*, meaning box. The capsid packages the viral genetic material/genome, shielding it from degradative enzymes.[7] Additionally, it recognizes host-cell receptors to facilitate infection and transport the genetic material to the replication site. *Capsid* refers to all the proteins that comprise the virus coat. The capsid is metastable, meaning that the capsomers are neither too strongly nor too weakly bound to each other. This characteristic maintains structural stability extracellularly and facilitates disassembly intracellularly to release viral genetic material.

Expressed viral proteins are produced within the host to facilitate viral replication. They are also referred to as non-structural proteins, indicating that they are not virion components. These proteins have functions that are essential for viral replication. Examples include proteins that form replication complexes (replicons) with host proteins, proteins that regulate the immune response of infected cells, and proteins that control the transcription rate or gene expression.

1.4 Lifecycle of Viruses

Cell entry is essential for infection and requires cell attachment. Subsequently, the virus replicates its genetic material and components, ultimately completing its lifecycle by exiting the infected cell. The viral lifecycle begins with infiltration of a host and entry to a target cell type. First, a free-floating virus attaches to and enters a cell. Viral entry can be divided into attachment, entry, movement, and uncoating steps.

Viral attachment occurs randomly, because viruses have no will or driving force. Upon encountering a host, a virus attaches to a suitable cell. Attachment factors facilitate viral binding to the cell surface, but they have low specificity. Therefore, there are few attachment factors that respond sensitively to specific viruses. By contrast, receptors have high specificity for particular viruses. The virus attaches to receptor proteins and begins cell entry.[8] As a result, the virus has enhanced affinity for cells that express those receptors. For example, HIV has high specificity for CD4, which explains its strong affinity for $CD4^+$ T cells. These specificities can enable the prediction of the infection potential of viruses by computational methods. Most receptors are proteins. However, vesicular stomatitis virus (VSV)

[7] In most cases, the capsid safely transports the viral genetic material to the replication site. However, there are instances, such as rotavirus, where the internal space of the capsid is used as the replication site.

[8] The preference of a virus for specific receptors does not mean that those receptors were designed for the virus. Rather, this preference is likely a result of adaptation by the virus. Viral infections are harmful to both the cell and the host organism. The mechanisms underlying the preference for certain receptors is a focus of research.

uses phosphatidylserine—a plasma-membrane component—as a receptor. VSV replicates its genetic material in the cytoplasm rather than in the nucleus. VSV is a negative-sense RNA virus of the family Rhabdoviridae. Some viruses enter cells by binding to specific receptors, while others directly attach to the cell membrane. Bacteriophages are an example of the latter, as are some viruses that cause human infections.

A consistent entry process among viruses would be a blessing for virus researchers; however, each viral species has a unique lifestyle. Although the processes vary slightly among viruses, the variations are not important enough to warrant detailed discussion. The processes are grouped into seven categories, which are known as the Baltimore groups.

Next, the virus releases its genetic material intracellularly (uncoating, gene expression, replication, assembly). Viral genetic material is replicated by various pathways. This is not a conscious or intentional action; it is based on thermodynamic or chemical reactions. The replication method differs depending on the type of virus. This is where the Baltimore classification system is useful. The progeny viruses replicate intracellularly and break out to infect new cells (release, maturation). If viruses had intelligence, this would be a mechanism to control, as random cell destruction and infection would lead to the annihilation of the host and, consequently, loss of their habitat. Ironically, although viruses lack consciousness, such mechanisms exist in nature. In other words, viruses that exhibit rapid (intra-host) infection and high mortality rates tend to be eliminated. This is similar to the development of herd immunity by host populations, which prevents transmission to other hosts. The next section focuses on the coronavirus. Given its effect on human health, coronavirus is a useful model for understanding viral behavior.

1.5 Coronavirus

Although it is essential to cover as many viruses as possible for completeness, summarising even a few important viruses is hampered by time and space constraints. As an alternative, I focus on the coronavirus that caused the recent pandemic, which is one of the most familiar viruses. In doing, so I address not only the virus itself but also public-health issues related to prevention measures. I hope that readers will use the insights gained from exploring and discussing the coronavirus in their investigations of other viruses of interest.

Coronaviruses (CoVs) cause respiratory and gastrointestinal infections in livestock and pets, and they account for approximately 15% of mild respiratory illnesses—including the common cold—in humans. The first CoV was identified 70 years ago. The severe acute respiratory syndrome coronavirus (SARS-CoV) was discovered in 2003, followed by the Middle East respiratory syndrome coronavirus (MERS-CoV) in 2012, highlighting the importance of research on CoVs responsible for severe acute respiratory diseases (Perlman and Netland 2009). The research stimulated by the SARS and MERS outbreaks led to advances in virology, genetics, phylogenetics, epidemiology, biology, pharmacology, and medicine. This research provided insights into CoV taxonomy, structure, lifecycle, genomic sequences, replication mechanisms, and transmission pathways, together with viral targets for the development of vaccines to prevent human CoV infections (Kwon et al. 2016).

In late 2019, a new CoV (SARS-CoV-2) emerged in Wuhan, China, triggering the COVID-19 worldwide pandemic, which resulted in an unprecedented public health crisis and social disruption. The pandemic demonstrated the importance of research and preparedness against the rapid mutations and spread of CoVs. SARS-CoV-2 continues to mutate, and the emergence of new variants with increased transmissibility and potential vaccine evasion has underscored the necessity of continuous surveillance and response strategies. In this context, examining the characteristics of CoVs, significant historical outbreaks, and the unique features of the COVID-19 pandemic is crucial. This will elucidate the ongoing and evolving threats these viruses pose to global health and emphasize the need for bioinformatics approaches to study and respond to them.

1.5.1 Key Outbreaks of Coronaviruses and the COVID-19 Pandemic

The first documented outbreak caused by coronaviruses was described in 1931 by Schalk and Hawn (Schalk and Hawn 1931). They identified the avian infectious bronchitis virus (IBV), which was confirmed by Beaudette and Hudson in 1937, as the cause of respiratory disease in birds (Beaudette and Hudson 1937). In the 1960s, two human coronaviruses were identified: the alphacoronavirus HCoV-229E and the betacoronavirus HCoV-OC43, both of which cause mild respiratory symptoms similar to the common cold in humans. They can also cause severe infections in immunocompromised individuals, infants, and the elderly (Reed 1984; Pyrc et al. 2007).

At the end of 2002, a new coronavirus that emerged in Guangdong, China, caused severe pneumonia and rapidly spread worldwide. The unusual pneumonia symptoms exhibited by infected patients showed characteristics of diffuse alveolar damage (DAD), which can progress to acute respiratory distress syndrome (ARDS). The virus was named SARS-CoV and was confirmed to have been transmitted from bats to humans via intermediate hosts such as Himalayan palm civets and raccoons (Guan et al. 2003; Li et al. 2006; Shi and Hu 2008). By 2003, SARS-CoV had caused 8098 infections and 774 deaths globally (Peiris et al. 2004), with an average fatality rate of about 9%. SARS-CoV infection was particularly severe in individuals over 60 years of age (fatality rate 50%) (World Health Organization 2003).

MERS-CoV was discovered in the Middle East in 2012. MERS-CoV caused a highly pathogenic respiratory disease with a fatality rate of 35%. Most MERS-CoV-infected patients had pre-existing chronic conditions such as diabetes or renal failure (Watson et al. 2014; Al-Tawfiq and Memish 2014). MERS-CoV is primarily transmitted to humans by dromedary camels, while human-to-human transmission is possible (Kayali and Peiris 2015). According to the World Health Organization (WHO), 1368 persons were infected with MERS-CoV between 2012 and 2015, resulting in 487 deaths (World Health Organization 2015). In 2015, a superspreading event occurred in South Korea, leading to the infection of 186 individuals with MERS-CoV in a short time (Durai et al. 2015).

In December 2019, SARS-CoV-2 was discovered in Wuhan, China. SARS-CoV-2 had genetic characteristics similar to SARS-CoV and had very high transmissibility. SARS-CoV-2, which causes COVID-19, spread rapidly worldwide, and the WHO declared a pandemic on March 11, 2020. The COVID-19 pandemic was one of the most serious public health crises of the twenty-first century and had tremendous social, economic, and cultural effects.

COVID-19 is highly transmissible between individuals, and asymptomatic carriers can spread the virus, hampering control measures. Initially, respiratory symptoms such as pneumonia, high fever, and cough are the primary clinical manifestations. However, COVID-19 also affects other organs, including those of the cardiovascular, nervous, and digestive systems, and has a range of clinical manifestations from asymptomatic to severe respiratory disease and multi-organ damage. Elderly individuals and those with underlying health conditions experienced severe disease, resulting in high mortality rates (Wu et al. 2020). Consequently, public health systems worldwide faced challenges such as shortages of medical resources, overwhelmed hospital capacities, and infections among healthcare workers.

In the early stages of the COVID-19 pandemic, the lack of effective treatments highlighted the importance of vaccine development as a key strategy for overcoming the crisis. Researchers worldwide quickly initiated vaccine development, and mRNA (Pfizer-BioNTech and Moderna) and viral-vector (AstraZeneca and Janssen) vaccines were developed within 1 year. This was the fastest vaccine development in history, and vaccination resulted in the gradual establishment of herd immunity.

The high mutation rate of SARS-CoV-2 resulted in the emergence of variant viruses (e.g., alpha, beta, gamma, delta, and omicron), which reduced vaccine efficacy. The delta and omicron variants, which emerged in 2021, had high transmissibility and immune-evasion characteristics, diminishing the effectiveness of control measures (Walls et al. 2020). The omicron variant has much higher transmissibility than other variants, with breakthrough infections occurring frequently among vaccinated individuals. This has underscored the necessity of booster shots and the development of vaccines effective against new variants (Nyberg et al. 2022).

The COVID-19 pandemic had a marked impact on the global economy and society. Governments worldwide implemented control measures such as lockdowns, social distancing, and mandatory mask-wearing to curb the spread of infections, resulting in a sharp contraction of economic activities and massive unemployment. Additionally, it brought about social changes such as the spread of non-contact culture, the normalization of remote work and education, and the rapid growth of e-commerce. The pandemic has significantly altered existing economic and social structures, and these changes are expected to persist after the pandemic.

1.5.2 Characteristics of CoVs

Coronaviruses are enveloped positive-sense RNA viruses of the family Coronaviridae, with a genome of 26 to 32 kb, making them the largest RNA viruses. Coronaviruses are classified into four genera (alpha, beta, gamma, and delta), and they infect a variety of animals (humans, mice, chickens, turkeys, swans, dogs, cats, rabbits, horses, and cows). The coronaviruses that infect humans include HCoV-229E, HCoV-OC43, HCoV-NL63, HCoV-HKU1, SARS-CoV, MERS-CoV, and SARS-CoV-2 (Milne-Price et al. 2014).

HCoV-HKU1, HCoV-NL63, HCoV-229E, and HCoV-OC43 cause mild respiratory infections characterized by upper respiratory symptoms such as the common cold, cough, and sore throat, which can lead to lower-respiratory-tract infections such as bronchitis, bronchiolitis, and pneumonia in immunocompromised individuals (Woo et al. 2005; Pyrc et al. 2007; Talbot et al. 2009). By contrast, SARS-CoV, MERS-CoV, and SARS-CoV-2 cause severe respiratory diseases and SARS-CoV-2 causes millions of deaths worldwide. These coronaviruses are transmitted primarily via respiratory droplets but also through contact and feces (Fehr and Perlman 2015).

CoV is an enveloped positive-sense RNA virus with a large RNA genome of 26–32 kb, with a 5′-cap and a 3′-poly(A) tail that encompasses at least six open reading frames (ORFs) (Lai and Cavanagh 1997). The CoV envelope consists of transmembrane (M) glycoprotein, spike (S) glycoprotein, and envelope (E) protein, which surround a helical nucleocapsid with symmetry (Neuman et al. 2006; Barcena et al. 2009). S protein is responsible for attaching to the host receptor and varies among viruses, making it the most important factor in host infection. M protein is the most abundant structural protein. E protein, although differing among species, has a common structure that facilitates viral assembly and release (Delmas and Laude 1990; Godet et al. 1992; Nal et al. 2005; Beniac et al. 2006; Nieto-Torres et al. 2014).

The CoV lifecycle begins with binding to the host-cell receptor. The CoV S protein binds to various host receptors, allowing the virus to penetrate the cytoplasm, a critical determinant of cell tropism. For example, SARS-CoV and SARS-CoV-2 use angiotensin-converting enzyme 2 (ACE2) as a receptor, whereas MERS-CoV uses dipeptidyl-peptidase 4 (DPP4). After receptor binding, the viral genome is released into the cytoplasm to initiate replication, and the replicated genome forms new viral particles for release.

The binding domain (RBD) in the S1 region of the S protein attaches to the host receptor, membrane fusion occurs via the S2 domain, facilitating entry, and the viral RNA genome is released into the cytoplasm. Next, the replicase gene is translated from the viral genome. The replicase gene comprises two large open reading frames (ORFs), rep1a and rep1b, which produce two final polyproteins, pp1a and pp1ab. To express these two polyproteins, the virus uses a slippery sequence, 5′-UUUAAAC-3′, and an RNA pseudoknot to induce ribosomal frameshifting from rep1a to rep1b (Brierley et al. 1989; Baranov et al. 2005). pp1a and pp1ab contain non-structural proteins (NSP) 1-11 and NSP 1-16, respectively, and are later cleaved into NSPs by one or two CoV proteases (Ziebuhr et al. 2000). These proteases include the papain-like protease (PLpro) encoded by NSP3 and the serine-type protease (main protease, Mpro) encoded by NSP5. Most CoVs encode two PLpros within NSP3, but SARS-CoV, MERS-CoV, and gammacoronaviruses encode only one PLpro (Mielech et al. 2014). The encoded proteases release components that are essential for assembling the viral replication complex to synthesize negative-strand RNA (Sethna et al. 1991).

Viral genomic RNA and sub-genomic RNA are next produced. Sub-genomic RNA serves as mRNA for structural and accessory genes, and in most cases, only the 5' ORF is translated. All positive-sense sub-genomic RNAs terminate at the same 3′, forming a nested RNA set characteristic of the order Nidovirales. Genomic RNA and sub-genomic RNA produce negative-strand intermediates that account for 1% of the positive-sense RNA and contain poly-uridylate and anti-leader sequences. During the synthesis of the negative strand, a discontinuous transcription strategy creates a nested set of negative-strand sub-genomic RNAs. The transcription regulatory sequence (TRS) at the 3′-end of the leading sequence is crucial. TRS models imply that RdRp can stop at a body TRS sequence (TRS-B). According to these models, when RdRp stops, it either continues extending to the next TRS or switches from TRS-B to the leader TRS to amplify the leader sequence at the 5′-end. This is supported by much evidence, including the presence of the anti-leader sequence at the 3′-end of negative-strand sub-genomic RNA (Sethna et al. 1991).

After replication and synthesis of sub-genomic RNA, the S, E, and M structural proteins are translated and inserted into the endoplasmic reticulum. These proteins are transported via the endoplasmic reticulum-Golgi intermediate compartment (ERGIC) pathway, and the viral genome is encapsulated by the N protein. The virus buds off from the ERGIC membrane containing the structural proteins, forming a mature virion (De Haan and Rottier 2005). Finally, the virion is transported to the cell surface and released by exocytosis.

CoVs with such a lifecycle replicate by mechanisms that differ from most other RNA viruses. CoVs have large genomes, and frequent errors during replication promote the exchange of genetic material between two or more CoVs in the same cell, resulting in the emergence of new variants with unpredictable host ranges and pathogenicities (Kahn 2006).

During the COVID-19 pandemic, the alpha, beta, gamma, delta, and omicron variants of SARS-CoV-2 emerged. These variants had increased transmissibility and immune-evasion capabilities. The delta variant was significantly more transmissible than previous variants, and the omicron variant had strong immune-evasion properties and rapidly became the dominant strain.

CoVs can spread by contact, via feces or respiratory droplets, and from infected animals to humans (zoonotic). Although respiratory droplets are the primary transmission route, CoVs can infect intestinal epithelial cells, potentially causing gastrointestinal diseases, and rarely, they can infect nervous tissue (Fehr and Perlman 2015). SARS-CoV-2 can affect several organs and tissues, leading to long COVID (Aiyegbusi et al. 2021).

The characteristics of CoVs significantly enhance their importance. Although several treatments for human-infecting CoVs have been approved, the early stages of the COVID-19 pandemic were marked by the absence of effective antiviral drugs or vaccines. Fortunately, vaccine development progressed rapidly, and emergency-use authorization of several vaccines reduced the rates of severe disease and mortality. However, breakthrough infections have been reported, highlighting the need for booster shots and the development of vaccines that can address emerging variants.

The most effective response to new CoVs is robust public health surveillance, rapid diagnostic testing, preventive measures, and, when necessary, isolation protocols. Additionally, during a global pandemic, collaboration and data sharing among governments, public health experts, and healthcare providers are essential (Fehr and Perlman 2015). This cooperation enables early detection of variant viruses and the establishment of effective response mechanisms.

Next-generation genomic sequencing and bioinformatics enable monitoring of the emergence of variant viruses in real time and accelerate the development of vaccines and treatments. Bioinformatics is important for analyzing and integrating genomic data, identifying new antiviral targets, and facilitating vaccine development, making it essential for responding to emerging infectious diseases. Based on this, it is important to establish a strategy that includes continuous evaluation of vaccine efficacy and the development of new treatments to prevent the emergence of new infectious diseases and respond effectively when they arise.

1.5.3 (Viro)Informatics Approaches in CoV Research

CoVs are zoonotic pathogens in various animal species and cause severe human disease. The outbreaks of SARS in 2003 and MERS in 2012 highlighted the need for research on CoVs and prompted research in phylogenetics, epidemiology, genetics, and virology. This research has provided insights into the structural features, replication mechanisms, lifecycles, and transmission pathways of CoVs and has identified viral targets for vaccine development.

The high mutation rate of CoVs leads to the emergence of new variants, potentially reducing the effectiveness of prevention and treatment strategies. The COVID-19 pandemic exacerbated these issues. SARS-CoV-2 has a very high mutation rate, and the alpha, beta, gamma, delta, and omicron variants emerged during the pandemic. These variants differ from the original virus in terms of transmissibility, pathogenicity, and immune-evasion ability, raising concern over the efficacies of vaccines and therapeutics.

In this context, bioinformatics is increasingly important for analyzing viral genetic characteristics, monitoring the emergence of variants, and providing essential information for the development of vaccines and therapeutics.

Publication of the genome sequence of SARS-CoV-2 in early 2020 prompted bioinformatics analyses of its structure, replication mechanism, and antigenicity (Zhu et al. 2020; Wu et al. 2020). The findings identified the key genetic regions of the virus and the factors influencing its infectivity and pathogenicity. Notably, structural mutations in the S protein were revealed to be determinants of receptor-binding affinity and antigenicity (Gobeil et al. 2021). Therefore, evaluation of vaccine efficacy and prediction of treatment effectiveness are essential because of the emergence of variants. Analysis of the mutation patterns of the S protein in the alpha, beta, gamma, delta, and omicron variants provided insight into their effects on transmissibility and immune evasion, which demonstrated the need for booster vaccinations (Carabelli et al. 2023).

Analysis of mutations in the NSPs of SARS-CoV-2 identified factors that enhance viral replication and strengthen immune evasion (Beyer and Forero 2022; Hossain et al. 2022). The results provided insights into the means by which the variants evade the host immune system and into the mechanisms of their pathogenicity. Comparative studies of MERS-CoV and SARS-CoV-2 sought to explain why the two viruses have significantly different transmissibility and pathogenicity despite their similar phylogenetic backgrounds (Rabaan et al. 2020). MERS-CoV had a high mortality rate but low person-to-person transmission, whereas SARS-CoV-2 had a low mortality rate but very high transmissibility. The results provided information on the differences in the structures, transmission routes, and immune-evasion mechanisms of the two viruses.

During the COVID-19 pandemic, databases and web-based tools were developed to track and analyze SARS-CoV-2 mutations. Platforms such as GISAID (Shu and McCauley 2017) and Nextstrain (Hadfield et al. 2018) contained genomic data on SARS-CoV-2, enabling monitoring and analysis of the emergence of variants in real time. These platforms contributed to tracking the transmission pathways of variants visually and predicting the spread of mutations.

Bioinformatics has facilitated the development of vaccines and therapeutics. Analysis of the structural characteristics and antigenic determinants of the SARS-CoV-2 S protein accelerated the development of mRNA vaccines (Wang et al. 2020; Jackson et al. 2020). The Pfizer-BioNTech and Moderna mRNA vaccines were designed based on bioinformatics analysis of the three-dimensional structure and immunogenicity of the S protein of SARS-CoV-2. This accelerated vaccine development and reduced the rates of severe disease and mortality.

Bioinformatics techniques enabled rapid assessment of the efficacy of vaccines and the establishment of new vaccine-development strategies tailored to new variants. Additionally, based on the structures of key SARS-CoV-2 proteins, such as 3CL protease and RNA-dependent RNA polymerase (RdRp), several antivirals targeting these proteins have been developed and are in clinical use, while others remain under development (Owen et al. 2021; Venkataraman et al. 2018). Virtual screening and molecular dynamics simulations have been used to predict the binding affinities of candidate drugs, thereby identifying optimal candidate compounds. For example, Remdesivir rapidly entered clinical trials after the prediction of its binding affinity for RdRp of SARS-CoV-2 (Beigel et al. 2020).

Bioinformatics approaches have been used in drug repurposing to evaluate the efficacy of existing drugs against SARS-CoV-2 rapidly. Virtual-screening and molecular-docking techniques have been used to assess the binding of approved drugs to SARS-CoV-2 proteins and to identify candidate antivirals (Elfiky 2020; Hosseini et al. 2021). However, contrary to expectations, drugs such as dexamethasone, hydroxychloroquine, and ivermectin did not exert significant therapeutic effects against SARS-CoV-2 in clinical trials, leading to controversy (RECOVERY Collaborative Group 2021; Self et al. 2020; Reis et al. 2022). This highlights the limitations of repurposing existing drugs, emphasizing the need for scientific verification of efficacy prior to the development of therapeutics.

Bioinformatics approaches are essential in CoV research, including in genomic analysis, monitoring of variant viruses, and the development of vaccines, diagnostics, and therapeutics. To maximize the potential of bioinformatics in a pandemic, international collaboration and data sharing are imperative. A system needs to be established for real-time sharing of genetic data on new viruses worldwide, analyzing the characteristics of variant viruses based on these data, and rapidly formulating countermeasures. Additionally, real-time analysis and integration of genomic data are essential to respond to the emergence of variant viruses. This will require advances in machine learning and artificial intelligence (AI). The use of AI for virtual screening of drug candidates, predicting protein-drug interactions, and identifying candidate vaccine antigens is accelerating the development of vaccines and therapeutics. For example, AI-based predictive models can simulate the transmission pathways of variant viruses and suggest prevention and treatment strategies. These technological advances will be vital in future pandemic responses. As demonstrated by the COVID-19 pandemic, genetic analysis and bioinformatics are important components of the response to epidemics. In the future, rapid genomic analysis, variant detection, and vaccine and treatment development using bioinformatics techniques will be important in the response to emerging infectious diseases. Understanding the genetic characteristics of viruses, predicting structural changes, and monitoring the emergence of variant viruses in real time will enable swift and effective responses to future pandemics.

1.5.4 Implications of Coronavirus Research

CoVs are important pathogens with unique genetic characteristics and high mutation rates, which humanity must combat. The outbreaks of SARS, MERS, and COVID-19 demonstrated the seriousness of the threat to public health posed by CoVs. These viruses rapidly mutate and can be transmitted from animals to humans, creating new pathogenic variants. Therefore, ongoing surveillance and research are essential.

The threat posed by CoVs extends beyond their pathogenicity; these viruses have the potential to undermine social, economic, and political stability. As the COVID-19 pandemic has shown, the global spread of a single virus can claim countless lives, paralyze the global economy, generate social unrest, and exacerbate political conflicts. Therefore, research on CoVs is important not only medically but also for ensuring sustainable development.

Due to the high mutation rates of pathogens (including CoVs), real-time monitoring and analysis are crucial for predicting and preventing pandemics. Bioinformatics enables rapid analysis of viral genomic changes, early detection of emerging variants, and assessment of the effects of mutations on the efficacy of vaccines and therapeutics. This can lead to more rapid and precise response strategies and the development of vaccines and therapeutics to control the spread of infectious diseases.

First, future CoV research and infectious-disease responses should focus on international collaboration and data sharing as the cornerstone of pandemic responses. The COVID-19 pandemic showed that no country is isolated in a connected world. Therefore, it is vital to monitor and share data on the emergence and spread of variant viruses in real time and develop joint responses on a global scale. This requires international agreements and governance systems that facilitate data sharing, together with close collaboration among governments and research institutions.

Second, the scope and depth of bioinformatics research should be expanded. Studies typically focus on genomic analysis, but future research should also encompass virus-host interactions, immune-evasion mechanisms, and environmental adaptation. Such research can enhance our understanding of the causes and consequences of mutations, going beyond merely tracking viral changes to predicting the long-term evolutionary paths of viruses.

Third, the application of bioinformatics research combined with artificial intelligence (AI) and machine learning (ML) should be maximized. AI and ML enable large-scale data analysis and predictive modeling, enabling real-time monitoring of genetic changes, prediction of the transmission pathways and pathogenicities of variants, and acceleration of the development of vaccines and therapeutics.

Finally, it is essential to develop sustainable infectious disease response strategies. Beyond short-term crisis management, it is necessary to strengthen long-term preparedness and prevention, expand the healthcare infrastructure and workforce, and continuously invest in research and development. This will enable humanity to respond more quickly and effectively to emerging infectious diseases.

CoVs are a persistent threat to human health. Addressing this threat requires scientific research, technological advances, international collaboration, and integrated responses. Bioinformatics approaches can connect these elements and protect human health in the face of emerging infectious diseases. With ongoing effort, we can confront the threats posed by novel viruses, minimize the impacts of pandemics, and ensure a sustainable future for humanity.

References

Adams MJ, Lefkowitz EJ, King AMQ, Harrach B, Harrison RL, Knowles NJ, Kropinski AM, Krupovic M, Kuhn JH, Mushegian AR, Nibert ML, Sabanadzovic S, Sanfaçon H, Siddell SG, Simmonds P, Varsani A, Zerbini FM, Orton RJ, Smith DB, Gorbalenya AE, Davison AJ. 50 years of the International Committee on Taxonomy of Viruses: progress and prospects. Arch Virol. 2017;162:1441–6.

Aiyegbusi OL, Hughes SE, Turner G, Rivera SC, McMullan C, Chandan JS, Haroon S, Price G, Davies EH, Nirantharakumar K, Sapey E, Calvert MJ. Symptoms, complications and management of long COVID: a review. J R Soc Med. 2021;114:428–42.

Al-Tawfiq JA, Memish ZA. Middle East respiratory syndrome coronavirus: transmission and phylogenetic evolution. Trends Microbiol. 2014;22:573–82.

Baranov PV, Henderson CM, Anderson CB, Gesteland RF, Atkins JF, Howard MT. Programmed ribosomal frameshifting in decoding the SARS-CoV genome. Virology. 2005;332:498–510.

Barcena M, Oostergetel GT, Bartelink W, Faas FG, Verkleij A, Rottier PJ, Bosch BJ. Cryo-electron tomography of mouse hepatitis virus: insights into

the structure of the coronavirion. Proc Natl Acad Sci USA. 2009;106:582–7.
Beaudette FR, Hudson CB. Cultivation of the virus of infectious bronchitis. J Am Vet Med Assoc. 1937;90:51–60.
Beigel JH, Tomashek KM, Dodd LE, Mehta AK, Zingman BS, Kalil AC, Hohmann E, Chu HY, Luetkemeyer A, Kline S, de Castilla DL, Finberg RW, Dierberg K, Tapson V, Hsieh L, Patterson TF, Paredes R, Sweeney DA, Short WR, Touloumi G, Lye DC, Norio Ohmagari N, Oh M, Ruiz-Palacios GM, Benfield T, Fätkenheuer G, Kortepeter MG, Atmar RL, Creech CB, Lundgren J, Babiker AG, Pett S, Neaton JD, Burgess TH, Bonnett T, Green M, Makowski M, Osinusi A, Nayak S, Lane HC. Remdesivir for the treatment of Covid-19. N Engl J Med. 2020;383:1813–26.
Beniac DR, Andonov A, Grudeski E, Booth TF. Architecture of the SARS coronavirus prefusion spike. Nat Struct Mol Biol. 2006;13:751–2.
Beyer DK, Forero A. Mechanisms of antiviral immune evasion of SARS-CoV-2. J Mol Biol. 2022;434:167265.
Brierley I, Digard P, Inglis SC. Characterization of an efficient coronavirus ribosomal frameshifting signal: requirement for an RNA pseudoknot. Cell. 1989;57:537–47.
Carabelli AM, Peacock TP, Thorne LG, Harvey WT, Hughes J, Peacock SJ, Barclay WS, de Silva TI, Towers GJ, Robertson DL. SARS-CoV-2 variant biology: immune escape, transmission and fitness. Nat Rev Microbiol. 2023;21:162–77.
Carroll KC, Pfaller MA, Landry ML, McAdam AJ, Patel R, Richter SS, Warnock DW, editors. Manual of clinical microbiology. Wiley; 2019.
De Haan CA, Rottier PJ. Molecular interactions in the assembly of coronaviruses. Adv Virus Res. 2005;64:165–230.
Delmas B, Laude H. Assembly of coronavirus spike protein into trimers and its role in epitope expression. J Virol. 1990;64:5367–75.
Durai P, Batool M, Shah M, Choi S. Middle East respiratory syndrome coronavirus: transmission, virology and therapeutic targeting to aid in outbreak control. Exp Mol Med. 2015;47:1–10.
Elfiky AA. Anti-HCV, nucleotide inhibitors, repurposing against COVID-19. Life Sci. 2020;248:117477.
Fehr AR, Perlman S. Coronaviruses: an overview of their replication and pathogenesis. Methods Mol Biol. 2015;1282:1–23.
Gobeil SMC, Janowska K, McDowell S, Mansouri K, Parks R, Stalls V, Kopp MF, Manne K, Li D, Wiehe K, Saunders KO, Edwards RJ, Korber B, Haynes BF, Henderson R, Acharya P. Effect of natural mutations of SARS-CoV-2 on spike structure, conformation, and antigenicity. Science. 2021;373:eabi6226.
Godet M, L'Haridon R, Vautherot JF, Laude H. TGEV coronavirus ORF4 encodes a membrane protein that is incorporated into virions. Virology. 1992;188:666–75.
Guan Y, Zheng BJ, He YQ, Liu XL, Zhuang ZX, Cheung CL, Butt KM. Isolation and characterization of viruses related to the SARS coronavirus from animals in southern China. Science. 2003;302:276–8.
Hadfield J, Megill C, Bell SM, Huddleston J, Potter B, Callender C, Sagulenko P, Bedford T, Neher RA. Nextstrain: real-time tracking of pathogen evolution. Bioinformatics. 2018;34:4121–3.
Hershey A, Chase M. Independent functions of viral protein and nucleic acid in growth of bacteriophage. J Gen Physiol. 1952;36:39–56.
Hossain A, Akter S, Rashid AA, Khair S, Alam ARU. Unique mutations in SARS-CoV-2 Omicron subvariants' non-spike proteins: potential impacts on viral pathogenesis and host immune evasion. Microb Pathog. 2022;170:105699.
Hosseini M, Chen W, Xiao D, Wang C. Computational molecular docking and virtual screening revealed promising SARS-CoV-2 drugs. Precis Clin Med e. 2021;4:1–16.
Jackson LA, Anderson EJ, Rouphael NG, Roberts PC, Makhene M, Coler RN, McCullough MP, Chappell JD, Denison MR, Stevens LJ, Pruijssers AJ, McDermott A, Flach B, Doria-Rose NA, Corbett KS, Morabito KM, O'Dell S, Schmidt SD, Swanson PA II, Padilla M, Mascola JR, Neuzil KM, Bennett H, Sun W, Peters E, Makowski M, Albert J, Cross K, Buchanan W, Pikaart-Tautges R, Ledgerwood JE, Graham BS, Beigel JH. An mRNA vaccine against SARS-CoV-2—preliminary report. N Engl J Med. 2020;383:1920–31.
Kahn JS. The widening scope of coronaviruses. Curr Opin Pediatr. 2006;18:42–7.
Kayali G, Peiris M. A more detailed picture of the epidemiology of Middle East respiratory syndrome coronavirus. Lancet Infect Dis. 2015;15:495–502.
King AM, Adams MJ, Carstens EB, Lefkowitz EJ. Virus taxonomy. Ninth Report of the International Committee on Taxonomy of Viruses, 9; 2012.
Kwon E, Cho M, Son HS. Bioinformatics for coronavirus research. JCBB. 2016;1:27–35.
Lai MM, Cavanagh D. The molecular biology of coronaviruses. Adv Virus Res. 1997;48:1–100.
Lapage SP, Sneath PH, Lessel EF, Skerman VBD, Seeliger HPR, Clark WA. International Code of Nomenclature of Bacteria and Statutes of the International Committee on Systematic Bacteriology and Statutes of the Bacteriology Section of the International Association of Microbiological Societies (Bacteriological Code, 1976, Revision, pp. xxxv+–180, 1975).
Le Gall O, Christian P, Fauquet CM, King AM, Knowles NJ, Nakashima N, Gorbalenya AE. Picornavirales, a proposed order of positive-sense single-stranded RNA viruses with a pseudo-T= 3 virion architecture. Arch Virol. 2008;153:715–27.
Li W, Wong SK, Li F, Kuhn JH, Huang IC, Choe H, Farzan M. Animal origins of the severe acute respiratory syndrome coronavirus: insight from ACE2–S protein interactions. J Virol. 2006;80:4211–20.
Linné CV. Systema Naturae, 1735: Facsimile of the First Edition: With an Introduction and a First English Translation of the "Observationes" No Title, 1735.

Mielech AM, Chen Y, Mesecar AD, Baker SC. Nidovirus papain-like proteases: multifunctional enzymes with protease, deubiquitinating and deISGylating activities. Virus Res. 2014;194:184–90.

Milne-Price S, Miazgowicz KL, Munster VJ. The emergence of the Middle East respiratory syndrome coronavirus. Pathog Dis. 2014;71:121–36.

Nal B, Chan C, Kien F, Siu L, Tse J, Chu K, van der Werf S. Differential maturation and subcellular localization of severe acute respiratory syndrome coronavirus surface proteins S, M and E. J Gen Virol. 2005;86:1423–57.

Neuman BW, Adair BD, Yoshioka C, Quispe JD, Orca G, Kuhn P, Buchmeier MJ. Supramolecular architecture of severe acute respiratory syndrome coronavirus revealed by electron cryomicroscopy. J Virol. 2006;80:7918–46.

Nieto-Torres JL, DeDiego ML, Verdia-Baguena C, Jimenez-Guardeno JM, Regla-Nava JA, Fernandez-Delgado R, Castaño-Rodriguez C, Alcaraz A, Torres J, Aguilella VM, Enjuanes L. Severe acute respiratory syndrome coronavirus envelope protein ion channel activity promotes virus fitness and pathogenesis. PLoS Pathog. 2014;10:e1004077.

Nyberg T, Ferguson NM, Nash SG, Webster HH, Flaxman S, Andrews N, Hinsley W, Bernal JL, Kall M, Bhatt S, Blomquist P, Zaidi A, Volz E, Aziz NA, Harman K, Funk S, Abbott S, COVID-19 Genomics UK (COG-UK) consortium, Hope R, Charlett A, Chand M, Ghani AC, Seaman SR, Dabrera G, Angelis DD, Presanis AM, Thelwall S. Comparative analysis of the risks of hospitalisation and death associated with SARS-CoV-2 omicron (B.1.1.529) and delta (B.1.617.2) variants in England: a cohort study. The Lancet. 2022;399:1303–12.

Owen DR, Allerton CM, Anderson AS, Aschenbrenner L, Avery M, Berritt S, Boras B, Cardin RD, Carlo A, Coffman KJ, Dantonio A, Di L, Eng H, Ferre R, Gajiwala KS, Gibson SA, Greasley SE, Hurst BL, Kadar EP, Kalgutkar AS, Lee JC, Lee J, Liu W, Mason SW, Noell S, Novak JJ, Obach RS, Ogilvie K, Patel NC, Pettersson M, Rai DK, Reese MR, Sammons MF, Sathish JG, Singh RSP, Steppan CM, Stewart AE, Tuttle JB, Updyke L, Verhoest PR, Wei L, Yang Q, Zhu Y. An oral SARS-CoV-2 Mpro inhibitor clinical candidate for the treatment of COVID-19. Science. 2021;374:1586–93.

Peiris JSM, Guan Y, Yuen KY. Severe acute respiratory syndrome. Nat Med. 2004;10:S88–97.

Perlman S, Netland J. Coronaviruses post-SARS: update on replication and pathogenesis. Nat Rev Microbiol. 2009;7:439–89.

Polaszek A. A universal register for animal names. Nature. 2005;437:477–7.

Pyrc K, Berkhout B, Van Der Hoek L. Identification of new human coronaviruses. Expert Rev Anti-Infect Ther. 2007;5:245–98.

Rabaan AA, Al-Ahmed SH, Haque S, Sah R, Tiwari R, Malik YS, Dhama K, Yatoo MI, Bonilla-Aldana DK, Rodriguez-Morales AJ. SARS-CoV-2, SARS-CoV, and MERS-COV: a comparative overview. Infez Med. 2020;28:174–84.

RECOVERY Collaborative Group. Dexamethasone in hospitalized patients with Covid-19. N Engl J Med. 2021;384:693–704.

Reed SE. The behaviour of recent isolates of human respiratory coronavirus in vitro and in volunteers: evidence of heterogeneity among 229E related strains. J Med Virol. 1984;13:179–92.

Reis G, Silva EA, Silva DC, Thabane L, Milagres AC, Ferreira TS, dos Santos CVQ, Campos VHS, Nogueira AMR, de Almeida APFG, Callegari ED, Neto ADF, Savassi LCM, Simplicio MIC, Ribeiro LB, Oliveira R, Harari O, Forrest JI, Ruton H, Sprague S, McKay P, Guo CM, Rowland-Yeo K, Guyatt GH, Boulware DR, Rayner CR, Mills EJ. Effect of early treatment with ivermectin among patients with Covid-19. N Engl J Med. 2022;386:1721–31.

Ride WDJL. International code of zoological nomenclature. International Trust for Zoological Nomenclature; 1999.

Schalk AF, Hawn MC. An apparently new respiratory disease of baby chicks. J Am Vet Med Assoc. 1931;78:413–22.

Self WH, Semler MW, Leither LM, Casey JD, Angus DC, Brower RG, Chang SY, Collins SP, Eppensteiner JC, Filbin MR, Files DC, Gibbs KW, Ginde AA, Gong MN, Harrell FE Jr, Hayden DL, Hough CL, Johnson NJ, Khan A, Lindsell CJ, Matthay MA, Marc Moss M, Park PK, Rice TW, Robinson BRH, Schoenfeld DA, Shapiro NI, Steingrub JS, Ulysse CA, Weissman A, Yealy DM, Thompson BT, Brown SM. Effect of hydroxychloroquine on clinical status at 14 days in hospitalized patients with COVID-19: a randomized clinical trial. JAMA. 2020;324:2165–76.

Sethna PB, Hofmann MA, Brian DA. Minus-strand copies of replicating coronavirus mRNAs contain antileaders. J Virol. 1991;65:320–5.

Shi Z, Hu Z. A review of studies on animal reservoirs of the SARS coronavirus. Virus Res. 2008;133:74–87.

Shu Y, McCauley J. GISAID: global initiative on sharing all influenza data–from vision to reality. Eur Secur. 2017;22:30494.

Talbot HKB, Crowe JE, Edwards KM, Griffin MR, Zhu Y, Weinberg GA, Williams JV. Coronavirus infection and hospitalizations for acute respiratory illness in young children. J Med Virol. 2009;81:853–9.

Uversky V, Longhi S. Flexible Viruses: Structural Disorder in Viral Proteins. Wiley; 2011.

Venkataraman S, Prasad BV, Selvarajan R. RNA dependent RNA polymerases: insights from structure, function and evolution. Viruses. 2018;10:76.

Walls AC, Park YJ, Tortorici MA, Wall A, McGuire AT, Veesler D. Structure, function, and antigenicity of the SARS-CoV-2 spike glycoprotein. Cell. 2020;181:281–92.

Wang F, Kream RM, Stefano GB. An evidence based perspective on mRNA-SARS-CoV-2 vaccine development. Med Sci Monit Int Med J Exp Clin Res. 2020;26:e924700–1.

Watson JT, Hall AJ, Erdman DD, Swerdlow DL, Gerber SI. Unraveling the mysteries of middle east respiratory syndrome Coronavirus. Emerg Infect Dis. 2014;20:1054–60.

Woo PC, Lau SK, Chu CM, Chan KH, Tsoi HW, Huang Y, Poon LL. Characterization and complete genome sequence of a novel coronavirus, coronavirus HKU1, from patients with pneumonia. J Virol. 2005;79:884–979.

World Health Organization. Consensus document on the epidemiology of severe acute respiratory syndrome (SARS), 2003.

World Health Organization. Investigation of cases of human infection with Middle East respiratory syndrome coronavirus (MERS-CoV): Interim guidance, 2015.

Wu A, Peng Y, Huang B, Ding X, Wang X, Niu P, Meng J, Zhu Z, Zhang Z, Wang J, Sheng J, Quan L, Xia Z, Tan W, Cheng G, Jiang T. Genome composition and divergence of the novel coronavirus (2019-nCoV) originating in China. Cell Host Microbe. 2020;27:325–8.

Zerbini FM, Siddell SG, Mushegian AR, Walker PJ, Lefkowitz EJ, Adriaenssens EM. Differentiating between viruses and virus species by writing their names correctly. Arch Virol. 2022;167:1231–4.

Zhu N, Zhang D, Wang W, Li X, Yang B, Song J, Zhao X, Huang B, Shi W, Lu R, Niu P, Zhan F, Ma X, Wang D, Xu W, Wu G, Gao GF, Tan W. A novel coronavirus from patients with pneumonia in China, 2019. N Engl J Med. 2020;382:727–33.

Ziebuhr J, Snijder EJ, Gorbalenya AE. Virus-encoded proteinases and proteolytic processing in the Nidovirales. J Gen Virol. 2000;81:853–932.

2 Bioinformatics in Brief

While the prefix "bio-" implies that bioinformatics is a discipline of biology, the specific field to which it belongs is not of particular importance. What's important instead is the recognition of bioinformatics as a valuable tool for various fields, especially those connected to biology. An example is public health, increasingly regarded as a vital applied science, which embraces various methodologies with the goal of "enhancing human health for the pursuit of happiness." Historically, public health has incorporated diverse academic approaches, a range recently growing further with the inclusion of bioinformatics. Public health is defined as a discipline rather than a science because its scope not only encompasses scientific methods but also aspects of the humanities and social sciences, reflecting its goal-oriented nature.

Before introducing widely used methods and theories in bioinformatics, it's important to establish an overall understanding of the field itself. There is a significant need to grasp basic concepts, terminology, and the general flow of the discipline before learning specific topics and techniques presented later. Therefore, this chapter will focus on a general introduction to bioinformatics, rather than detailing every aspect. While these basic concepts provide important foundations, they may not directly apply to practical work or research. More practical concepts and techniques will be introduced in the following sections and chapters.

2.1 Overview of Bioinformatics

Historically, bioinformatics began to gain significant attention following the successful completion of the Human Genome Project in 2003. This accomplishment brought recognition of the potential of these tools across various academic fields. Initially, methods of bioinformatics were perceived as narrow genetic approaches for determining genetic information, specifically DNA sequences. However, researchers in computational biology recognized the close links between many aspects that the two fields seek to explore and explain. This belief evolved, allowing bioinformatics to expand beyond its initial narrow definition into broader and more diverse areas.

While bioinformatics and computational biology are often discussed as distinct fields, they are not mutually exclusive. In fact, they share many common goals, methodologies, and techniques. For the purposes of this book, I will consider them to be closely related, if not equivalent. Both fields utilize computational approaches to analyze biological data, and the distinction between them is often more a matter of historical context and focus rather than a fundamental difference in methodology. Over time, the lines between these fields have blurred, and in many contexts, the terms can be used interchangeably.

Additionally, although the theories and methods historically referred to as bioinformatics have not significantly changed, many have been newly

H. S. Son, M. Cho, *Virinformatics and Computational Virology in Public Health*,
https://doi.org/10.1007/978-981-96-5757-5_2

defined within different research contexts, leading to terms such as Biomedical Informatics and Medical Informatics. In my personal view, for a new academic name to have merit, it must have intrinsic necessity; otherwise, it leads to unnecessary confusion and potential misdirection, especially for students new to the scientific field.

I intend therefore to approach bioinformatics in a broad sense, acknowledging that—similar to public health—it lacks clearly defined boundaries, making it difficult to state what should or shouldn't be included. This applies universally to all applied sciences, including bioinformatics. Some argue that bioinformatics possesses inherent foundational characteristics, and it is merely a matter of time before it is recognized as a fundamental science. While this view has its merits, in the context of this book, I expect that readers will view bioinformatics predominantly from an applied perspective, as it overwhelmingly possesses applied elements.

Now, let's explore the fundamental reasons behind the redefinition and rising prominence of bioinformatics. Although it may feel like a distant event, the Human Genome Project, conducted in the early twenty-first century, stands as a historical landmark for several reasons. Firstly, there is the practical issue of processing such vast amounts of data. Modern biology inherently benefits from collecting and organizing large datasets. Just classifying ecological data can constitute a specialized field due to its magnitude. Furthermore, advances in molecular biology and genetics have compounded this by adding large quantities of genetic information, rendering classical storage methods, like bibliographic forms, inadequate.

Secondly, data generated in individual laboratories must be integrated and analyzed collaboratively. Finite resources mean it is impractical for an individual lab to explore everything about a single species; thus, data must be part of collective scientific efforts to have value beyond individual researchers' projects.

Thirdly, developments in hardware, software, and the internet have reached a technological level necessary to meet these needs. Meanwhile, the extensive data collection has increased the demand for computational techniques to facilitate analysis. While reading and understanding numerous library books could theoretically grant access to most of humanity's knowledge, this would be impractical due to time constraints. Therefore, identifying and applying necessary and appropriate knowledge has become crucial. The same applies to data stored on computers; simply accumulating it without analysis is not productive. Thus, technology that aids experts in making informed decisions or enhancing understanding is essential and can be mediated by computational devices and programs. Techniques commonly referred to as data mining encompass these processes, enabling the practical and efficient use of the knowledge amassed by humanity.

2.2 The Human Genome Project

The term "genome" refers to the complete set of genetic information of an organism, encompassing the entirety of its DNA. A portmanteau of "gene" and "chromosome," it was first used by German scientist Hans Winkler in 1920. The human body consists of approximately 100 trillion cells, each containing 22 pairs of autosomes and a single pair of sex chromosomes. These chromosomes are composed of double-stranded DNA, about 2 meters in length, coiled around histone proteins. The goal of the Human Genome Project was to analyze the DNA sequences of all 46 chromosomes, to determine the locations and functions of their genes. Today, individual genomic information can be obtained quickly and at low cost, as a result of extensive research into the genetic material that underlies biological phenomena. It has been established that DNA, rather than protein or RNA, is the fundamental unit of life, leading to the continuous accumulation of associated knowledge. Following Watson and Crick's revelation of the double-helix structure of DNA in 1953, research began to determine DNA sequences (Watson and Crick 1953). In 1977, the Sanger sequencing method allowed for the reading of DNA fragment base-pair sequences (Sanger et al. 1977). The Maxam–

Gilbert method was another sequencing technique developed in this time. The development of the Polymerase Chain Reaction (PCR) method in 1983, which enables the mass amplification of genes, brought researchers a step closer to genome analysis (Mullis and Faloona 1987). In 1985, the first meeting aimed at interpreting the human genome was convened, marking the beginning of discussions on arranging the 3 billion base pairs and identifying gene locations.

On October 1, 1990, the U.S. Department of Energy (DOE) and the National Institutes of Health (NIH) launched the Human Genome Project (HGP) consortium, with the goal of analyzing human DNA and creating a genetic map within 15 years; this map would be available to all scientists. The initiative involved an investment of $3 billion and participation from over 350 research institutes across 18 countries outside the U.S. Meanwhile, Dr. Craig Venter founded the private company Celera Genomics in 1992, announcing his intention to complete the Human Genome Project by 2001, independently from the U.S. government. The resulting competition between the HGP consortium and Celera Genomics contributed to the earlier-than-expected completion of the genome analysis.

Alongside the human genome, the project has also revealed the genomes of over 20 diverse organisms, including nematodes, the bacterium *E. coli*, and fruit flies. The genome analysis of *Haemophilus influenzae* was completed in 1995, making it the first genome of a living organism to be decoded (Fleischmann et al. 1995). Following this, *Caenorhabditis elegans* was the first multicellular organism genome sequenced in 1998. In March 2000, Celera Genomics successfully determined the genome sequence of the fruit fly (Adams et al. 2000).

The genome information construction method used by the HGP consortium was known as the clone contig method. This technique involved cutting DNA and arranging the fragments sequentially to create a chromosome map. The cut DNA fragments were inserted into Bacterial Artificial Chromosome (BAC) vectors and overlapping regions of the clones were merged to form clone contigs. Each contig's sequence was then determined using the shotgun sequencing method. By matching the sequences of each revealed clone to the marker positions on genetic and physical maps, the larger genome sequence was completed (International Human Genome Sequencing Consortium 2001). In contrast, Venter's proposed whole genome sequencing method obtained sequences corresponding to 6.5 to 8 times the total genome length, without creating a genetic map. This approach involved fragmenting the genome, immediately decoding the nucleotide sequences, and then connecting the contigs to determine the order of each piece. Venter demonstrated that using a supercomputer to combine numerous contigs—a method previously thought to be impossible—was indeed feasible.

In December 1999, the HGP consortium successfully decoded the 22nd of the 46 human chromosomes. Following this, in 2000, they obtained the sequence information of the 165 million base-pair genome of the fruit fly and then the sequence of the human 21st chromosome. Originally planned for 15 years, the Human Genome Project reached a significant milestone on June 26, 2000, when U.S. President Bill Clinton, alongside Dr. Francis Collins from the National Human Genome Research Institute and Dr. Craig Venter from Celera Genomics, officially announced the completion of the draft human genome map. Following the public release of the draft in February 2001, the HGP consortium published their analysis results in Nature, while Celera Genomics published theirs in Science. Ultimately, the HGP consortium succeeded in interpreting 99% of the entire human genome sequence with an accuracy of 99.9% by April 2003 (International Human Genome Sequencing Consortium 2004). Reviews of the papers published by the HGP consortium and Celera suggest that the data quality from Celera was viewed as having notable relative strengths. This is attributed to the fact that the HGP consortium, with its analysis divided among multiple institutions, encountered numerous errors and created a standard human genome sequence by stitching together genomes from different individuals.

The initial goal of the Human Genome Project was to identify genes and determine the 3 billion base pairs that make up human DNA. In addition, it aimed to compile this information into databases and improve analysis techniques. According to the HGP consortium, the expected number of genes was not 100,000, but rather around 20,000 to 25,000, with an average gene consisting of about 3000 base pairs. It was also found that 95% of the genome consists of DNA sequences with unknown functions, and 45% comprises repetitive sequences; functional genes account for only about 1% of the genome. Furthermore, they discovered that human chromosomes have a relatively high frequency of repetitive sequences at the centromeres and telomeres in comparison with other organisms. The proteins produced from human genes were discovered to have more complex structures than those produced by invertebrates, and mutation rates during meiosis were found to be approximately twice as high in males than in females. The human genome is 99.9% identical among individuals, with the 0.1% difference arising from the presence of single nucleotide polymorphisms (SNPs). These SNPs, which vary at specific locations among individuals, number around 3 to 5 million in the genome and are factors influencing traits such as skin, hair, eye color, and height.

The Human Genome Project is one of the greatest research achievements of the twentieth century, providing insights into the mysteries of the core processes of life. However, the exact locations and functions of all genes are yet to be fully elucidated, leaving unresolved challenges. Addressing these issues would pave the way for a significant leap toward an era of personalized medicine.

2.3 Scope of Bioinformatics

Imagine a biologist who possesses a complete laboratory setup to investigate biological phenomena and has discovered associated laws through experiments. This biologist inherently embodies two key attributes: that of a data collector and that of a logical reasoner. Every scientist, to some extent, relies on these two qualities to conduct scientific work. Typically, we differentiate between experimentalists and theorists based on which attribute they emphasize more or from which approach their significant discoveries arise. Just as no experimentalist can solve problems without logical thinking, there cannot be a bioinformatician who works without data. However, a bioinformatician can be seen as a theorist whose role as a logical reasoner is maximized.

Of course, nothing prevents a bioinformatician from conducting wet experiments of their own volition; in practice, however, it is rare for bioinformaticians to engage in hands-on experimentation, and one might question whether this is even advisable. In terms of roles, this is what characterizes bioinformaticians: their primary experiments are those conducted with datasets on computers to elucidate phenomena.

2.4 Definition of Bioinformatics

Attempts to define the field of bioinformatics often produce considerable confusion. Some consider it a subset of genetics or genomics, while others view it as a branch of proteomics; some suggest it encompasses both areas. While none of these definitions are necessarily incorrect, the author believes that excluding any particular field renders the definition of bioinformatics significantly incomplete. Given these considerations, various definitions exist, but here I define bioinformatics as follows: "Bioinformatics is the science of managing and analyzing biological data using advanced computing techniques." It is important to note that bioinformatics is not merely a simple combination of biology and informatics; many other fields contribute to addressing its challenges.

Bioinformatics can be understood as the integration of computational biology (CB) and information technology (IT), where CB encompasses biology, computing, and various interdisciplinary fields such as mathematics and statistics. This emphasizes the fact that bioinformatics fundamentally serves as a conceptual framework:

$$\text{Bioinformatics} = \text{CB} + \text{IT}$$

where

CB = biology + computer science + various interdisciplinary fields + mathematics, statistics, etc.

2.5 Key Fields in Bioinformatics

The data derived from studying biology and conducting research not only serve as the foundation for subsequent work but also provide both direct and indirect benefits to humanity. In research processes aimed at solving biological problems, key terms that one inevitably encounters include sequences, structures, functions, and networks. From the researcher's perspective, biological research can be understood as the process of transforming initial data, composed of sequences, into meaningful information. Creating or predicting useful outcomes in the real world based on this information falls within the realm of application. Almost without exception, living organisms construct physical structures from sequences (RNA or DNA) containing genetic information, and these structures express functions essential for sustaining life. As in the various fields of biology, bioinformatics is typically classified based on these fundamental elements.

Bioinformatics as a discipline has significantly broadened its scope over the years. There are numerous unresolved challenges in biology, most of which are diverse and complex. Recent advances in bioinformatics technology can substantially assist in addressing these challenges. Given the extensive range of fields covered by biology, it can be said that all these related areas may also fall under the domain of bioinformatics. Within this wide scope, there are two primary methods for classifying bioinformatics: methodological and subject-based.

2.6 Bioinformatics Workflow

The first step in solving a biological problem is for the problem itself to be identified and defined. Once this is done, the next essential elements are data and tools. Data serves as the starting point for bioinformatics. The meaningful products of processing and analyzing this data are referred to as information, and this information can be applied in various ways. Bioinformatics can thus be seen as a series of processes that generate and make use of such information. As noted earlier, data is typically obtained through experiments and maintained in storage systems but often requires proper processing. To process and analyze this data, specific methods are needed, usually formulated as algorithms. These methods are implemented through software tools. On the hardware side, computers are typically used.[1]

2.6.1 Biological Databases and Websites

Obtaining data as the result of an experiment has certain advantages for analysis, in that subsequent steps can be performed relatively independently. However, even in such cases, there are often instances where direct or indirect comparisons with public domain data are necessary. From a computer science perspective, a database is a systematically organized collection of data that allows for useful and efficient use. In biology, due to the advances in computational infrastructure and technology, almost all experimental data generated today is stored and managed using computers. As a result, handling data through computational techniques has become possible. Additionally, many of these bioinformatics databases are publicly available through the web for the benefit of the broader community. In many cases, these sites also provide or are integrated with tools that allow users to analyze the available

[1] In extreme cases, the human brain could serve this role. However, such cases are beyond the scope of this discussion, and bioinformaticians generally use computational equipment with appropriate performance as their tools.

data, which is of great help to biological researchers. The forms of these tools vary, and the methods of implementation are equally diverse; this can pose a challenge in terms of fully grasping the characteristics of the data and mastering their usage. However, thanks to the integration of databases and tools, these resources are becoming increasingly user-friendly.

Databases used in bioinformatics are numerous and vary according to the nature of the data. In theory, there could be an infinite number of types. Databases exist for almost every form of data, from DNA and proteins to large-scale dynamic networks. The volume of biological data has already surpassed what could be referred to simply as "big data." A tremendous effort is required to systematically classify and store the vast quantities of different types of data, with specialized fields emerging dedicated to this task. From the perspective of bioinformaticians, the goal is simply to conveniently retrieve the necessary data and achieve the intended analysis objective. Therefore, our main concern is how to appropriately use websites or databases available on the internet. Table 2.1 provides a list of representative sites and URLs. Since our primary interest is in viruses, there is no need to list the full range of biological databases; however, there are occasions when a broader perspective or related data is necessary for comparison. Please refer to this list with a broader view in mind.

Table 2.1 Biological databases and resources

Organization	Description	Website
NCBI	An NIH-affiliated agency providing databases and tools for research in genomics, bioinformatics, and biotechnology worldwide	https://www.ncbi.nlm.nih.gov/
Ensembl	Provides gene annotations, variant data, and gene expression information for various species, enabling researchers to explore and analyze data through a web-based interface and various tools	https://www.ensembl.org
UCSC Genome Browser	Provides tools and databases for visually exploring genomic sequence data of vertebrates, invertebrates, and major model organisms, including humans	https://genome.ucsc.edu/
1000 Genomes Project	A large-scale database analyzing human genomic variations. The reference data resources generated by this project are widely used in the biomedical scientific community	https://www.internationalgenome.org/
DDBJ	A Japanese nucleotide sequence database that collects and stores biological sequence information, primarily DNA sequence data and annotations, supporting researchers in searching and analyzing biological data	https://www.ddbj.nig.ac.jp/index-e.html
Gencode	A database providing gene annotations for humans and mice	https://www.gencodegenes.org/
UniProt	A comprehensive database for protein sequences and functions, covering protein function, structure, modifications, and interactions	https://www.uniprot.org/
RCSB PDB	A database for 3D structure data of large biological molecules, including proteins and nucleic acids, providing structural information obtained through X-ray crystallography, NMR, and electron microscopy	https://www.rcsb.org/
InterPro	A database that classifies proteins into families and provides domain and functional analysis data	https://www.ebi.ac.uk/interpro/
AlphaFold Protein Structure Database	Provides protein structures predicted by the AlphaFold model	https://alphafold.ebi.ac.uk/
PRIDE	A proteomics database that stores and provides experimental data related to proteins	https://www.ebi.ac.uk/pride/

(continued)

Table 2.1 (continued)

Organization	Description	Website
KEGG	A database that provides information on metabolic pathways, gene functions, and disease pathways and visualizes gene and molecular interactions in network form	https://www.genome.jp/kegg/
COSMIC	A database that collects information on somatic mutations related to cancer, allowing analysis of mutations in specific cancer types	https://cancer.sanger.ac.uk/cosmic
OMIM	A database that provides information on human genetic disorders and related genes	https://www.omim.org/
ExPASy	A platform offering various bioinformatics tools and resources, providing access to protein-related databases	https://www.expasy.org/
miRBase	A database for miRNA sequence and function information	https://www.mirbase.org/
RNAcentral	A database that integrates all known non-coding RNA sequences from various species	https://rnacentral.org/
ClinGen	A database that integrates and organizes clinical information on genes and genetic variants	https://clinicalgenome.org/
DECIPHER	Provides information on genetic and chromosomal abnormalities along with clinical data based on these findings	https://www.deciphergenomics.org/
PharmGKB	A pharmacogenomics database that provides information on drug responses and interactions with genes	https://www.pharmgkb.org/
TCGA	Provides genomic data, sequencing data, transcriptomic data, proteomic data, and methylation data for various cancer types	https://www.cancer.gov/ccg/research/genome-sequencing/tcga
STRING	Provides information on protein-protein interaction networks	https://string-db.org/
BioGRID	A database that collects and integrates various biological interaction data, supporting the understanding and study of the complexity of biological systems	https://thebiogrid.org/
IntAct	A database that provides data on biological interactions, particularly focusing on protein-protein interaction (PPI) data	https://www.ebi.ac.uk/intact/home
GTEx	Provides gene expression levels across human tissues and is useful for studying the correlation between gene expression and genotype	https://gtexportal.org/home
GeneCards	A comprehensive resource for human gene information, integrating data on genes, proteins, phenotypes, and related diseases	https://www.genecards.org/
Human Protein Atlas	Provides data analyzing protein expression patterns at the tissue and cellular level, useful for studying in vivo protein expression and localization	https://www.proteinatlas.org/

2.6.2 Bioinformatics Repositories

Collections of software tools represent other important repositories. Depending on our needs, we can download and use tools developed by fellow developers/researchers. Traditionally, there have been various large and small websites that have provided bioinformatics tools, but the most widely known platform in recent years is GitHub (Chacon and Straub 2014; Kalliamvakou et al. 2014).

GitHub is one of the world's largest platforms on which software developers can effectively collaborate and share code, and it has become a central hub for open-source projects. As the volume of biological data grows exponentially, the development and utilization of software tools to efficiently process and analyze such data is essential. Exploring bioinformatics-related open-source projects, tools, and libraries on GitHub can be very useful for learning to handle this data. By studying the code of open-source projects, one can learn about the structure and methodologies of code used in actual research, as well as gain insight into current research trends by reviewing the tools in use. Additionally, anyone can contribute to the community by creating and sharing their own projects.

There are several ways to find bioinformatics-related software or tools on GitHub. The search function is a helpful starting point. By typing keywords such as "bioinformatics," "genomics," "sequence alignment software," "Python bioinformatics," or "R bioinformatics" into the search bar at the top, and applying various filters on the search results page, you can narrow results down to match your needs. You can also use the "Topics" feature, which tags projects with specific subjects, to locate relevant information. You can either enter the topic into the search bar or search for repositories by entering the URL in the format "github.com/topics/topic-name." Furthermore, you can explore "Awesome Lists," which are collections of useful software, libraries, and tools for specific subjects or fields. Searching for projects such as "awesome-bioinformatics" or "awesome-genomics" can help you find resources in your area of interest.

If you want to develop and share bioinformatics tools on GitHub, here's how to do it. First, create an account on GitHub and create a new repository. You will need to set the project name, description, and license and, if needed, choose between public or private settings to control access. To write the code, choose a widely used programming language in the field—such as Python, Java, R, Perl, or C++—and develop the tool's code in your local development environment. During coding, use Git for version control to track changes. Commit your work regularly and use branches to separate tasks such as new features or bug fixes. Write a README file that includes instructions on usage, installation, and examples. Select an appropriate open-source license for the project and include it in the repository. Before distributing the code, test the tool to ensure it works as expected, and if necessary, write test cases. Once the code is complete, push it to the GitHub repository to share it. GitHub's "Releases" feature allows you to distribute specific versions of the software and, if needed, include installation files or binaries.

Some representative GitHub repositories in bioinformatics include the following:

Biopython: A Python library for biological data processing that allows users to easily perform various tasks, such as sequence analysis, protein structure analysis, and genomic data processing
(https://github.com/biopython/biopython.git)

DeepVariant: A gene variant detection tool developed by Google that uses a deep learning-based approach to accurately and reliably detect variants, primarily for identifying single nucleotide variants and insertion-deletion mutations in Next Generation Sequencing (NGS) data
(https://github.com/google/deepvariant.git)

SeqKit: A sequence data analysis tool that efficiently performs common tasks on FASTA and FASTQ formatted files
(https://github.com/shenwei356/seqkit.git)

BWA: A tool for the fast and accurate alignment of DNA sequences to a reference genome, helping to match large-scale DNA sequencing data to genomic databases
(https://github.com/lh3/bwa.git)

Bioconda: A collection of software and tools used in various biological fields, based on the Conda package manager
(https://github.com/bioconda/bioconda-recipes.git)

Galaxy: An open-source, web-based platform for designing and executing analysis pipelines in bioinformatics and data science, commonly used in genomics and proteomics analysis
(https://github.com/galaxyproject/galaxy.git)

2.6.3 Application of Simulation Methods

Simulation refers to the use of a virtual experiment in place of a real-world study. The idea of a virtual experiment is quite intuitive, as it's common to attach modifiers to the word "experiment," with "thought experiments" being another typical example. It has now been quite some time since simulations, which differ from traditional practical experiments, first emerged. If we include the realm of thought experiments, which use the human brain as a simulation, simulations have been ongoing throughout the history of humankind. However, in the context of scientific simulations, the idea cannot be discussed without considering computers. With the rapid advances in computer-related technologies in the twentieth century, simulations have become an essential element in many fields and play a pivotal role in computational science.

Since our interest lies in biological research, simulation can be introduced as a subfield of computational biology. As previously noted, we will not strictly differentiate between bioinformatics and computational biology, so introducing simulation techniques and recommending their use is a natural extension of the discussed methods.

Simulations allow for the virtual implementation of experimental environments that would be difficult to carry out in reality. However, in many cases, while certain objectives have been achieved, it would be hard to say that they have been fully realized (at least in the case of computational biology). The field still requires many improvements in terms of detailed precision. Nevertheless, despite existing limitations, the achievements and future potential of simulations are vast. This is why simulation techniques cannot be overlooked in bioinformatics or computational biology. Section 2.7 presents a brief overview of simulations in biology, addressed independently as a distinct part of the text.

2.6.4 Developing Bioinformatics Software

A serious bioinformatician is likely to want to create new software based on their own innovative ideas. If that software becomes widely used, it brings not only recognition as a bioinformatician but also a deep sense of satisfaction. Many software tools have already been developed, following the hard work of previous researchers, and most problems can now be easily solved using these existing tools. Moreover, with the advancement of artificial intelligence (AI), it is undeniable that writing software has become much easier compared to the past. However, despite this accumulation of tools and advancements, I believe that the discovery of new methodologies based on human creativity will continue, alongside new software to implement them.

When attempting to write your own software, the first thing you should do is choose a programming language that you are comfortable with. While Python is the most popular choice, there is no harm in trying other languages. Once you become familiar with a language, the next task will be to define the problem and translate it into code. In this process, methods like system analysis or logical analysis may be helpful. However, the most important thing is to check if the software you want already exists in full or in part. It is never advisable to reinvent the wheel unless it is for practice or as a hobby.

Once the software is created, it is advisable to open it up for others to use. It is also a good idea to find a suitable repository for the software and, if appropriate, publish it as a paper.

2.7 Simulation Methods

Simulation can be defined as the process of reproducing or predicting temporal changes based on a model. Modeling, on the other hand, is the process of creating a representation of a real phenomenon and involves simplifying complex real-world systems. Despite the clear differences between them, simulation and modeling are undoubtedly intertwined and often used interchangeably.

In the field of biology, simulations are appropriately classified in relation to components of living organisms. The components of living organisms can be divided into organisms, organs, tissues, cells, and molecules. Individual organisms and their combinations form larger units such as populations, communities, ecosystems, and the biosphere. It is well known that simulation techniques can be applied to each of these categories. For example, simulations of disease transmission, a critical topic in public health, attempt to explain phenomena at the population and community levels using agent-based modeling. In this way, simulation techniques have not only contributed to the biological sciences but also to closely related fields like public health.

Although our ultimate focus is on public health related to viruses, it is beneficial to approach simulation techniques from a broader biological perspective; the importance of studying components such as proteins related to viruses cannot be overlooked. For example, simulations can be used to understand the adsorption specificity between a host's receptor and the spike protein. It is also possible to simulate evolutionary trends at the genomic level. Among various existing biological simulations, this section will introduce several simulation approaches focused on protein structure.

Simulation methods can be divided into "dynamic methods," such as molecular dynamics, and "static methods," such as energy minimization or Monte Carlo (MC) methods. The distinction between them is based on how the concept of time is applied. Simulations that do not incorporate time focus solely on determining the equilibrium state or probability distribution of a system. In contrast, simulations that incorporate the concept of time emphasize tracking dynamic temporal changes. Dynamic methods require the assignment of initial velocities to the particles that make up the system, whereas static methods do not require any information about the system's motion. In both methods, structural information is important. In dynamic methods, the primary description of the system is derived at each stage of the simulation, whereas in static methods, it is provided by the stable final structure.

Static methods have an obvious disadvantage compared to dynamic methods, in that they can only provide spatial information. Dynamic methods, in contrast, offer information about the time dependence and magnitude of fluctuations in both position and motion. However, in practice, both methods are often combined to simulate the system of interest. Additionally, the concept of "simulated annealing" (SA) can be applied to both methods and, when used with an MC search protocol, can also provide path information regarding structural changes (Kirkpatrick et al. 1983).

2.7.1 Potential Function

To apply such simulation techniques, a description of the system is required, which is achieved by constructing a potential function for the system. This function contains a set of quantitative parameters needed to evaluate the energy values; both the function and the parameter set constitute the "force field" of the system. In theory, this potential function can be constructed through quantum mechanical considerations. However, due to limitations related to the system size, an empirical Newtonian energy potential function is used in most cases. The energy functions used for biological systems such as proteins are typically composed of covalent terms representing bond lengths, angles, van der Waals interactions, and electrostatic contributions (McCammon and Harvey 1987).

In order to characterize the dynamics of a biological system, it is crucial to have a function that describes the potential energy in relation to the positions of atoms within the system. Although this potential can theoretically be derived from quantum mechanical principles, its application to large systems is impractical due to the complexity and limitations involved. As a result, classical potential functions have been developed as an alternative. While these classical potentials aim to encompass all possible interatomic interactions, practical limitations lead to the use of reasonable approximations and assumptions, resulting in the inclusion of only a subset of these interactions. A common representation of such a function for biological systems is as follows (McCammon and Harvey 1987):

$$V = \frac{1}{2}\sum_{\text{bonds}} K_b\left(b-b_o\right)^2 + \frac{1}{2}\sum_{\text{bond\_angles}} K_\theta\left(\theta-\theta_o\right)^2 + \frac{1}{2}\sum_{\text{dihedral\_angles}} K_\varphi\left[1+\cos\left(n\varphi-\delta\right)\right] + \sum_{\text{nonbonded\_pairs}}\left(\frac{q_1 q_2}{4\pi\varepsilon r} + \frac{A}{r^{12}} - \frac{B}{r^6}\right)$$

The first summand in the potential function represents the energy contribution from all covalent bonds within the molecule. Each bond is modeled according to Hooke's law, assuming a force constant K_b and an equilibrium bond length b_0. This approximation is valid under typical biological conditions, where bond length fluctuations are minimal. The second term accounts for the deformation energy of valence angles θ formed between covalent bonds attached to a given atom, with θ_o being the ideal bond angle for any particular triplet of bonded atoms. The third term describes the intrinsic energy associated with twisting covalently bonded atoms around an axis, where n represents periodicity, and δ is the reference angle with respect to ϕ. Together, these three terms capture the variations in covalent bonding energy within the molecule. The remaining term includes three sub-terms corresponding to electrostatic interaction, core repulsion, and dispersion between pairs of atoms separated by a distance, r (McCammon and Harvey 1987).

Additionally, an "improper torsion term" can be incorporated into the potential to maintain chirality and planarity, as expressed in the following equation:

$$K_\omega(\omega-\omega_o)^2$$

where K_ω is a constant and ω_o represents the ideal improper torsion angle. The ideal values of b_o, θ_o, and ω_o are derived from X-ray diffraction studies of small molecules, while the force constants K_b, K_θ, K_ϕ, and K_ω are based on infrared vibrational spectroscopy results (Brooks et al. 1988).

This potential function is a typical energy function used in Molecular Dynamics (MD) simulations in major simulation packages, such as AMBER (Weiner and Kollman 1981), CHARMm (Brooks et al. 1983), and Xplor (Brünger 1992; Schwieters et al. 2003). However, simpler forms of this potential function may be employed in other simulations. For instance, Milik and Skolnick studied the insertion of peptide chains into lipid membranes by simplifying each amino acid into a single peptide in their model. As a result, the internal energy (i.e., the potential function) was simplified to include only two terms: bond and torsion angle terms between residues (Milik and Skolnick 1993).

A complete description of the system requires specifying the parameters that characterize the individual terms of the potential energy function. These parameters must be provided separately for various atom types. Given that proteins contain a relatively small number of atom types compared to the total number of constituent atoms, using a set of parameters has proven to be an efficient method for organizing this information. The energy potential function, along with the parameter set, together form the "force field" of the system.

A potential energy function and its associated parameter set, also known as a force field, provide the foundation for studying the motion and structure of proteins.[2] The following section will focus on energy minimization and molecular dynamics methods.

2.7.2 Energy Minimization Method

Energy minimization in biological systems is a method used to identify a conformation with minimal energy, employing an unconstrained multivariate nonlinear optimization technique. All minimization methods of this type follow an iterative process. If the previous estimate of the minimum value of the function is X_{i-1}, the new estimate, X_i, is computed as follows:

$$X_i = X_{i-1} + \delta x V_{i-1}$$

Here, $\boldsymbol{V}_{i-1}$ represents the search direction determined from the previous point, $\boldsymbol{X}_{i-1}$. It is important to note that $\boldsymbol{X}_{i-1}$ and $\boldsymbol{X}_i$ are position vectors. The displacement along the direction $\boldsymbol{V}_{i-1}$ is denoted as δx. To determine the new $\boldsymbol{X}_i$, an appropriate search method to find both δx and $\boldsymbol{V}_{i-1}$ is required.

[2]The methods most frequently employed are molecular dynamics and energy minimisation. Molecular dynamics simulations, which solve Newton's equations of motion for a system's atoms and the surrounding solvent, offer the most detailed insights into the system's behaviour. Structural and dynamic properties can be obtained from simulations lasting just a few picoseconds.

Minimization techniques can be categorized into two main types: direct search methods and gradient-based methods. Direct search methods do not rely on partial derivatives of the function (and are often referred to as non-gradient methods). Examples of these methods include the Nelder–Mead simplex method (Nelder and Mead 1965) and the Hooke–Jeeves pattern search method (Hooke and Jeeves 1961). While these methods are suitable for problems with a relatively small number of variables, they tend to be less efficient than gradient-based methods. Gradient-based methods, on the other hand, require the first (and possibly higher) derivatives of the cost function—in this case, the empirical potential function—as well as evaluations of the function. "Steepest descent" and the "conjugate gradient method" are two of the most commonly used gradient optimization techniques. The steepest descent algorithm computes the direction of descent based on the gradient at the current potential. However, it cannot overcome local energy barriers because the gradient direction is the sole factor influencing structural movement. This makes it inefficient for multidimensional optimization problems that involve rugged energy surfaces with multiple local minima. In contrast, the conjugate gradient method is generally more efficient because it takes into account the gradients from both the current and previous steps to determine the direction.

2.7.3 Molecular Dynamics Simulation Method

MD is fundamentally a method for simulating the motion of a system composed of particles. This approach has played a crucial role in defining the motion of biological molecules such as proteins, nucleic acids, oligosaccharides, and lipids (McCammon and Harvey 1987; Singh et al. 1985; Harvey and Prabhakaran 1986; Pastor and Venable 1993). The motion of a molecule is integral to its functional role. For example, hemoglobin's binding of oxygen requires structural fluctuations (Karplus and Pesko 1990).

When applied to proteins, knowledge of these motions can help understand various biological functions, such as ligand binding. MD simulations are used in structure determination and refinement through data from Nuclear Magnetic Resonance (NMR) or X-ray diffraction experiments (Brünger 1992). In X-ray experiments, the procedure for refining the structure typically involves performing MD after an initial energy minimization step to relieve local stresses in the structure.

In MD, the forces acting on the atoms of a system are calculated by differentiating the potential energy terms with respect to the atomic positions. The initial structure for the simulation can be derived, for example, through homology modeling (Hubbard and Lesk 1995). This is equivalent to obtaining X-ray crystallography or NMR data. MD simulations are also used to calculate the free energy changes resulting from mutations in proteins (Karplus and Pesko 1990).

MD is based on the calculation of numerical values related to the motion of particles. For these calculations, approaches such as the Verlet method (Nelder and Mead 1965), Beeman method (Beeman 1976), fifth-order Gear method (Gear 1971), and the SHAKE method (Ryckaert et al. 1977; van Gunsteren and Berendsen 1977) have been developed. The widely used Verlet algorithm is derived from the Taylor series expansion and is summarized as follows:

$$R_i(t+\Delta t) = 2R_i(t) - R_i(t-\Delta t) + \frac{\Delta t^2}{m_i} f_i(t) + O(\Delta t^4)$$

where $\boldsymbol{R}_i(t)$ is the position of particle i at time t, $\boldsymbol{R}_i(t + \Delta t)$ represents the predicted position of particle i at the next time step, $t + \Delta t$, and $\boldsymbol{R}_i(t - \Delta t)$ is the position of particle i at the previous time step, $t - \Delta t$. Δt is the time step, which is the interval between each update in the simulation. m_i refers to the mass of particle i, $\boldsymbol{f}_i(t)$ is the force acting on particle i at time t, $O(\Delta t^4)$ is the error term, which indicates the higher-order approximation error that results from the time step Δt. From this equation, if the current position and the position at Δt prior are known, the position at Δt ahead can be predicted.

In most cases, MD simulations can be performed using existing software. Numerous MD-related software packages have likely been developed and are currently in use. Below are a few well-known software tools that are commonly used in biological MD simulations.

GROMACS: A high-performance software that supports various MD simulations, including those for biomolecules, polymers, and physicochemical systems. It is available as free, open-source software and offers fast computation speeds through parallel computing (Berendsen et al. 1995).
(https://www.gromacs.org)

LAMMPS (Large-scale Atomic/Molecular Massively Parallel Simulator): An open-source software for a wide range of MD simulations, capable of simulating atomic interactions and molecular systems. It supports a variety of physical models and force fields (Thompson et al. 2022).
(https://www.lammps.org/)

AMBER: A software primarily used for simulating biomolecules, especially proteins, nucleic acids, and ligands. It combines MD and quantum chemistry calculations to provide accurate analyses (Case et al. 2023). (https://ambermd.org).

CHARMM (Chemistry at Harvard Macromolecular Mechanics): A software used for molecular modeling and simulations, primarily for studying the structure and dynamics of proteins, nucleic acids, and other biomolecules (Brooks et al. 1983). (https://www.charmm.org).

In recent years, the development of machine learning-based potentials (MLPs) has significantly enhanced the accuracy of force fields by enabling more precise predictions of molecular interactions. This has been demonstrated in multiple studies, such as the work of Kolloff and Olsson (2022), which explored how machine learning can improve MD simulations for biomolecular systems (Kolloff and Olsson 2022). Additionally, machine learning and artificial intelligence techniques are being integrated into MD simulations, allowing for more efficient conformational searches and more accurate modeling of biological systems (Kaptan and Vattulainen 2022).

The potential function could benefit from the inclusion of polarization effects or, ideally, from the use of a quantum mechanical potential. However, applying quantum mechanics to an entire system remains impractical due to the large number of atoms involved. The electron count in a protein molecule often exceeds the limits of current computational power, and solving the Schrödinger equation for all possible interactions is prohibitively time-consuming, even for smaller systems. Nevertheless, recent advances in hybrid quantum mechanics/molecular mechanics (QM/MM) methods have made it possible to incorporate quantum mechanical calculations for specific regions of the system, enhancing accuracy without the computational cost of a full quantum treatment (Senn and Thiel 2009).

Moreover, new developments in quantum computing have the potential to revolutionize molecular simulations by providing the computational power needed to solve quantum mechanical problems at a much larger scale. Although this is still in its early stages, the future integration of quantum computers into MD simulations could significantly expand the scope of simulations that can be performed with high accuracy.

In addition to these challenges, MD simulations are constrained by the lengths of simulation time required, which are influenced by both the system size and the time step used in calculations. Conventional MD simulations typically employ small time steps of around 0.5–1.0 fs; this means even simulations covering short time spans require extensive computational resources. However, recent advances in GPU acceleration and parallel computing have made it possible to perform long-timescale simulations more efficiently. Enhanced sampling techniques, such as Replica Exchange MD, have also been developed to improve the exploration of complex energy landscapes, making it possible to study systems over longer timescales with reduced computational expense (Zhou et al. 2011; Bernardi et al. 2015).

Several strategies have been proposed to address these limitations. For example, quantum mechanical calculations can be integrated into classical MD simulations for regions where more precise information is required (Senn and Thiel 2009). Furthermore, the use of high-throughput simulations has become more widespread, enabling large-scale screenings of molecular interactions. These approaches are particularly useful in drug discovery and other applications that require the exploration of vast chemical spaces (Durrant and McCammon 2011; Aldewachi et al. 2021).

In summary, MD simulations remain an essential tool for studying biological macromolecules, and recent developments in force field design, quantum mechanical methods, and computational power are helping to overcome many of the method's traditional limitations. The integration of artificial intelligence, machine learning, and quantum computing is poised to further enhance the capability and efficiency of MD simulations in the years to come.

2.7.4 Simulated Annealing and the Monte Carlo Simulation Method

2.7.4.1 Theory and Methods

The principle behind SA is derived from a well-known physical phenomenon. When a metal is heated to its melting point and then rapidly cooled, it becomes brittle, indicating that its atoms are locked into a local minimum energy state, leading to an unstable structure. However, if the metal is cooled slowly, it becomes much more stable, suggesting that it has formed the most stable arrangement, corresponding to the global minimum energy. Similarly, SA works by initially heating the system to a high temperature, enabling it to overcome energy barriers between local minima and the global minimum. The system is then slowly cooled, encouraging it to settle into lower-energy configurations. If the force field is accurate, the final conformation obtained at low temperatures will ideally correspond to the lowest possible energy state of the system.

SA is frequently combined with MD and MC methods to optimize molecular systems. The MC method was originally developed by von Neumann and Ulam to model neutron diffusion and was named after the random number generation used. The method was later extended by Metropolis and others to include the concept of probabilistic acceptance in the simulation process. In broad terms, MC is a random search technique and relies on trial and error.

SA is a global optimization technique that draws its inspiration from the Metropolis algorithm (Metropolis et al. 1953), which established a probabilistic criterion for accepting new configurations in a system based on energy differences. This foundational concept was later expanded upon by Kirkpatrick et al. (1983), who introduced the concept of gradually lowering the temperature to allow the system to explore a range of configurations, enabling it to escape local minima and eventually settle into the global minimum energy configuration. This development led to the creation of SA as a method specifically designed to optimize global energy minimization in complex optimization problems (Kirkpatrick et al. 1983).

2.7.4.2 The Metropolis Prescription

To apply the SA method for energy minimization, a temperature-dependent selection mechanism is integrated into the simulation. Metropolis introduced a random selection approach (the MC search) to implement the SA concept. The Metropolis prescription (Metropolis et al. 1953) decides whether to accept each proposed structure based on the Boltzmann probability, defined as:

$$P = e^{\Delta E(q)/kT}.$$

where $\Delta E(q)$ is the energy difference between the new structure $\boldsymbol{q}_{i+1}$ and the previous structure $\boldsymbol{q}_i$, $\boldsymbol{k}$ is the Boltzmann constant, and $\boldsymbol{T}$ is the absolute temperature.

When energy minimization is applied to a system, such as a protein molecule, the initial structure is typically derived from experimental techniques such as X-ray crystallography or computational modeling. The simulation then involves randomly selecting new structures and calculating their potential energy. If the new structure has lower energy than the previous one, it is accepted as the current minimized structure. This process continues until the energy converges.

The Metropolis prescription differs from the basic MC energy minimization method in that it permits the random selection of higher-energy structures. As described by the equation, if the energy of the new structure is higher than the current structure, it will still be accepted based on the Boltzmann probability, *P*. The decision to accept or reject a higher-energy structure is made by comparing the Boltzmann probability *P* with a random number ξ, which is uniformly distributed between 0 and 1. The comparison works as follows:

Acceptance of higher-energy structures: If ($\boldsymbol{P} \geq \boldsymbol{\xi}$), the new structure is accepted, even if it has higher energy. This allows the system to escape local minima, particularly in the early stages of the simulation when the system is at a high temperature and needs to explore a larger conformational space.

Rejection of higher-energy structures: If ($\boldsymbol{P} < \boldsymbol{\xi}$), the new structure is rejected, ensuring that at lower temperatures, the system converges to the global minimum rather than getting trapped in local minima.

This mechanism allows for uphill movements at high temperatures, which become less likely as the temperature decreases. The Boltzmann probability determines the likelihood of accepting higher-energy configurations, and as the system cools down, it becomes less likely to accept such configurations. For each temperature, $\boldsymbol{N}$ steps of conformational search are performed. The search space and the number of steps, *N*, should be defined prior to running the simulation.

2.8 Bioinformatics and Viroinformatics in Public Health

This chapter provides a brief introduction to bioinformatics. Bioinformatics and computational biology are recognized as essential tools and methodologies for solving biological problems. When these methodologies are combined with the unique characteristics and challenges of viruses, fields such as viroinformatics and computational virology emerge to play an indispensable role in public health.

Infectious diseases represent a critical issue in public health. Having recently experienced a global crisis caused by a novel virus, we are filled with hope that informatics tools will effectively address future threats. For instance, it is no longer an unrealistic goal to construct sophisticated simulations to predict viral mutations and use these to design more effective vaccines. Building systems capable of processing the vast amount of viral data, which is incomparable to general biological data, is another significant but no longer insurmountable challenge.

A crucial point here is the central role that the development of AI will play in the future of bioinformatics and viroinformatics. AI can be an invaluable tool for analyzing large volumes of biological data and detecting patterns in viral mutations. For example, machine learning algorithms can be used to uncover hidden patterns in existing data, allowing for more precise predictions of viral mutations. This could provide crucial information for proactively responding to future threats.

AI could also make significant contributions to the development of vaccines and treatments. By rapidly analyzing data and predicting candidate therapeutic substances, the discovery process can be made much more efficient and effective. Advances in AI will drive significant innovation in public health through collaboration with viroinformatics and computational virology. This synergy will bring about essential change to infectious disease response strategies and lay the groundwork for effectively addressing future health threats.

References

Adams MD, et al. The genome sequence of *Drosophila melanogaster*. Science. 2000;287:2185–95.

Aldewachi H, Al-Zidan RN, Conner MT, Salman MM. High-throughput screening platforms in the discovery of novel drugs for neurodegenerative diseases. Bioengineering. 2021;8:30.

Beeman D. Some multistep methods for use in molecular dynamics calculations. J Biochem. 1976;54:43.

Berendsen HJC, van der Spoel D, van Drunen R. GROMACS: a message-passing parallel molecular dynamics implementation. Comput Phys Commun. 1995;91:43–56.

Bernardi RC, Melo MCR, Schulten K. Enhanced sampling techniques in molecular dynamics simulations of biological systems. Biochim Biophys Acta. 2015;1850:872–7.

Brooks BR, Bruccoleri RE, Olafson BD, States DJ, Swaminathan S, Karplus M. CHARMM: a program for macromolecular energy minimisation, and dynamics calculations. J Comp Chem. 1983;4:187–217.

Brooks CL, Karplus M, Pettitt BM. Proteins: a theoretical perspective of dynamics, structure and thermodynamics. Wiley; 1988.

Brünger AT. X-PLOR version 3.0. A system for X-ray crystallography and NMR. Yale University Press; 1992.

Case DA, Aktulga HM, Belfon K, Cerutti DS, Cisneros GA, Cruz Eiro VWD, Forouzesh N, Giese TJ, Götz AW, Gohlke H, Izadi S, Kasavajhala K, Kaymak MC, King E, Kurtzman T, Lee T-S, Li P, Liu J, Luchko T, Luo R, Manathunga M, Machado MR, Nguyen HM, O'Hearn KA, Onufriev AV, Pan F, Pantano S, Qi R, Rahnamoun A, Risheh A, Schott-Verdugo S, Shajan A, Swails J, Wang J, Wei H, Wu X, Wu Y, Zhang S, Zhao S, Zhu Q, Cheatham TE III, Roe DR, Roitberg A, Simmerling C, York DM, Nagan MC, Merz KM Jr. AmberTools. J Chem Inf Model. 2023;63:6183–91.

Chacon S, Straub B. Pro Git. 2nd ed. Apress; 2014.

Durrant JD, McCammon JA. Molecular dynamics simulations and drug discovery. BMC Biol. 2011;9:71.

Fleischmann RD, et al. Whole-genome random sequencing and assembly of *Haemophilus influenzae* Rd. Science. 1995;269:496–512.

Gear CW. Numerical initial value problems in ordinary differential equations. New York: Prentice-Hall; 1971.

Harvey SC, Prabhakaran M. Ribose puckering: structure dynamics, energetics and the pseudorotation cycle. J Am Chem Soc. 1986;108:6128–36.

Hooke R, Jeeves TA. Direct search solution of numerical and statistical problems. J Assn Comput Mach. 1961;8:212–29.

Hubbard TJP, Lesk AM. Computer modelling in molecular biology: modelling protein structures. New York: VCH Publishers; 1995. p. 9–35.

International Human Genome Sequencing Consortium. Initial sequencing and analysis of the human genome. Nature. 2001;409:860–921.

International Human Genome Sequencing Consortium. Finishing the euchromatic sequence of the human genome. Nature. 2004;431:931–45.

Kalliamvakou E, Gousios G, Blincoe K. The promises and perils of mining GitHub. In: Proceedings of the 11th working conference on mining software repositories (MSR '14); 2014. p. 148–57.

Kaptan S, Vattulainen I. Machine learning in the analysis of biomolecular simulations. Adv Phys X. 2022;7:2006080.

Karplus M, Pesko GA. Molecular dynamics simulations in biology. Nature. 1990;347:631–9.

Kirkpatrick S, Gellatt CD Jr, Vecchi MP. Optimization by simulated annealing. Science. 1983;220:671–80.

Kolloff C, Olsson S. Machine learning in molecular dynamics simulations of biomolecular systems. arXiv:2205.03135 [q-bio.BM], 2022.

McCammon JA, Harvey SC. Dynamics of proteins and nucleic acids. Cambridge University Press; 1987. p. 39–44.

Metropolis N, Rosenbluth AW, Rosenbluth MN, Teller AH. Equation of state calculations by fast computing machines. J Chem Phys. 1953;21:1087–92.

Milik M, Skolnick J. Insertion of peptide chains into lipid membranes: an off lattice Monte Carlo Dynamics model. Proteins. 1993;15:10–25.

Mullis K, Faloona F. Specific synthesis of DNA in vitro via a polymerase-catalyzed chain reaction. Methods Enzymol. 1987;155:335–50.

Nelder JA, Mead RA. A simplex method for function minimisation. Comput J. 1965;7:308–13.

Pastor RW, Venable RM. Computer simulation of biomolecular systems: theoretical and experimental applications. ESCOM Sci Publishers BV. 1993;2:443–63.

Ryckaert JP, Ciccotti G, Berendsen HJC. Numerical integration of the Cartesian equations of motion of a system with constraints: molecular dynamics of n-alkanes. J Comput Phys. 1977;23:327.

Sanger F, Nicklen S, Coulson AR. DNA sequencing with chain-terminating inhibitors. Proc Natl Acad Sci USA. 1977;74:5463–7.

Schwieters CD, Kuszewski JJ, Tjandra N, Clore GM. The Xplor-NIH NMR molecular structure determination package. J Magn Reson. 2003;160:65–73.

Senn HM, Thiel W. QM/MM methods for biomolecular systems. Angew Chem Int Ed. 2009;48:1198–229.

Singh UC, Weiner SJ, Kollman P. Molecular dynamics simulation of d(C-G-C-G-A)d(T-C-G-C-G) with and without "hydrated" counterions. Proc Natl Acad Sci USA. 1985;82:755–9.

Thompson AP, Aktulga HM, Berger R, Bolintineanu DS, Brown WM, Crozier PS, In 't Veld P J, Kohlmeyer A, Moore SG, Nguyen TD, Shan R, Stevens MJ, Tranchida J, Trott C, Plimpton SJ. LAMMPS – a flexible simulation tool for particle-based materials modeling at the atomic, meso, and continuum scales. Comput Phys Commun. 2022;271:10817.

van Gunsteren WF, Berendsen HJC. Algorithms for macromolecular dynamics and constraint dynamics. Mol Phys. 1977;34:1311–27.

Watson JD, Crick FHC. Molecular structure of nucleic acids: A structure for deoxyribose nucleic acid. Nature. 1953;171:737–8.

Weiner PK, Kollman PA. AMBER: assisted model building with energy refinement. A general program for modeling molecules and their interactions. J Comp Chem. 1981;2:287–303.

Zhou Y, Liepe J, Sheng X, Stumpf MPH, Barnes C. GPU accelerated biochemical network simulation. Bioinformatics. 2011;27:874–6.

Part II

Essential Methods for Viroinformatics

3 Phylogenetics

Molecular phylogenetics investigates the evolutionary relationships among molecules such as genes and proteins by analyzing their structure, function, and changes over time. This field emerged in the early twentieth century and gained significant momentum in the 1960s following advancements in molecular biology techniques, including gene and protein sequencing, PCR, and gel electrophoresis (Durbin et al. 1998; Dowell 2008). Since then, increased accessibility to computers and the sophistication of algorithms have allowed for precise definitions and explanations of molecular evolution through complex probabilistic and computational methods (Lio and Godman 1998).

Molecular phylogenetics primarily aims to reconstruct the sequence of evolutionary events and visually represent relationships among species or genes using evolutionary trees (phylogenies). This process is nuanced and multifaceted (Linder and Warnow 2001). Originally focused on systematic and taxonomic studies, phylogenetics now encompasses nearly every branch of biology. It elucidates relationships among species as well as paralogous genes within gene families, the evolutionary history of populations, and the adaptation of pathogens (Mäser et al. 2001; Marra et al. 2003; Grenfell et al. 2004; Edwards 2009).

Recent advancements in whole-genome sequencing have broadened the scope of molecular phylogenetics, making it an essential tool for comparing genomes across diverse organisms, from humans to microbes (Yang and Rannala 2012). As a result, molecular phylogenetics addresses a wide range of research topics, including the classification of genome sequences; the elucidation of genes, regulatory elements, and non-coding RNAs within newly acquired genome sequences; and the interpretation of both ancestral and contemporary genomic information (Kellis et al. 2003; Pedersen et al. 2006; Paten et al. 2008; Green et al. 2010; Gronau et al. 2011; Lindblad-Toh et al. 2011; Li and Durbin 2011; Ma 2011).

In this chapter, we explore the foundational concepts, methodologies, and analytical stages of molecular phylogenetics. We also examine its diverse applications in evolutionary analysis and discuss future directions for the field. As a notable example, we highlight phylogenetic analysis in tracing the evolutionary relationships among primate lentiviruses, such as HIV-1, HIV-2, and SIV, illustrating how molecular phylogenetics facilitates the understanding of cross-species transmission and zoonotic origins. Through these discussions, we aim to provide a comprehensive understanding of molecular phylogenetics, highlighting its crucial role in advancing biological sciences and research on viruses.

H. S. Son, M. Cho, *Viroinformatics and Computational Virology in Public Health*,
https://doi.org/10.1007/978-981-96-5757-5_3

3.1 Understanding Phylogenetic Analysis

3.1.1 Basic Concepts of Tree Construction

A phylogenetic tree, which resembles a branching tree structure, represents the evolutionary relationships among genetic lineages over time. Each branch signifies the continuity of these lineages, while each node marks the emergence of new lineages. When a phylogenetic tree illustrates relationships among species, the nodes typically represent speciation events, although interpretations can vary. For example, in a gene tree based on sequences from a group, nodes can be seen as the inferred ancestors of the sampled individuals. In contrast, in a tree of paralogous gene families, the nodes indicate gene duplication events (Yang and Rannala 2012; Cho and Son 2017).

As shown in Fig. 3.1, branches and nodes can be located internally within the tree or at its external tips, with terminal nodes representing operational taxonomic units (OTUs). OTUs correspond to molecular sequences or taxa (species) within the inferred tree, while internal nodes represent the last common ancestor (LCA) of all descendant nodes. Phylogenetic trees can be based on a single gene derived from multiple taxa or multigene families (Baldauf 2003; Hall 2004). A rooted tree features a specific node or outgroup from which all OTUs are derived, establishing a root—the oldest point in the tree representing the common ancestor of all taxa. When no confirmed outgroup is available, the root may be placed centrally, or an unrooted tree can be constructed.

Branches in phylogenetic trees can be categorized into groups based on distinct characteristics. Figure 3.2 illustrates monophyletic groups, which include an internal LCA node and all OTUs derived from it, which share a unique set of inherited traits. Paraphyletic groups, such as "all mammals except for members of the order Primates," exclude some descendants. Polyphyletic groups, on the other hand, exhibit similar traits or phenotypes that suggest relatedness but do not share a common ancestor, including OTUs from more distantly related groups (Li 1997; Baldauf 2003).

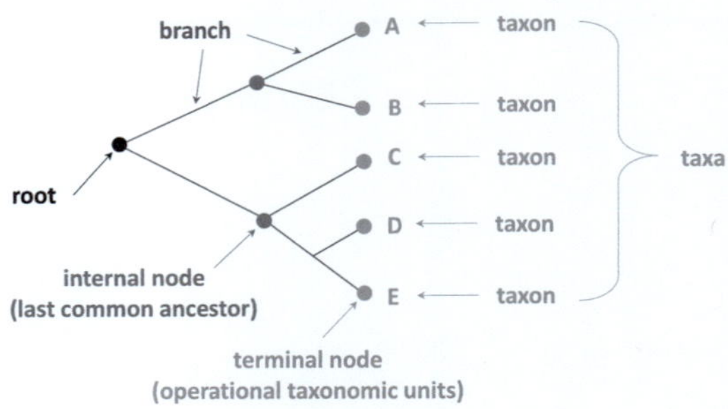

Fig. 3.1 Basic components of a phylogenetic tree

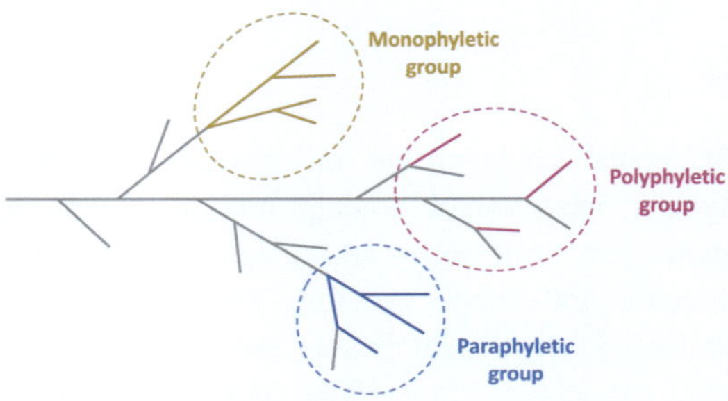

Fig. 3.2 Taxonomic groups in a phylogenetic tree

Phylogenetic trees are scientific reconstructions of molecular data, such as DNA or protein sequences, rather than direct observations. The process of creating them, known as "phylogenetic reconstruction," employs a range of algorithms and computational methods tailored to the data characteristics and research goals. These methods are broadly categorized into distance-based and character-based approaches (Felsenstein 2004).

Distance-based methods calculate pairwise distances between sequences to create a distance matrix, which serves as the foundation for tree construction (Fitch and Margoliash 1967). For example, the neighbor-joining (NJ) method utilizes a clustering algorithm on this matrix (Saitou and Nei 1987).

Character-based methods include maximum parsimony (MP), maximum likelihood (ML), and Bayesian inference (BI) (Felsenstein 2004). These approaches assess the fit of each tree by evaluating specific positions (characters) across aligned sequences. MP aims to minimize the number of substitutions, ML computes log-likelihood values, and BI determines posterior probabilities (Saitou and Imanishi 1989; Hillis et al. 1994; Takahashi and Nei 2000).

While comparing all possible trees would be ideal theoretically, this is computationally impractical for large datasets. Therefore, heuristic algorithms are employed to explore tree space (Hillis and Huelsenbeck 1992; Lewis 1998). These heuristic methods use rapid algorithms to generate initial trees and perform local rearrangements to refine tree scores, although they do not guarantee an optimal tree (Lewis 1998).

In terms of data interpretation, the distance matrix, ML, and BI methods are considered model-based as they use substitution models. In contrast, MP does not rely on a specific model but instead uses implicit assumptions for tree estimation (Yang and Rannala 1997, 2012; Lemmon et al. 2009).

3.1.2 Steps in Phylogenetic Analysis

The nature and scope of phylogenetic studies can vary widely, necessitating different datasets and computational methods. However, fundamentally, they generally involve three key stages: "dataset construction," "phylogenetic tree reconstruction using computational methods and probabilistic models," and "statistical validation and evaluation of inferred trees" (Hall 2004).

3.1.2.1 Dataset Construction (Data Collection and Alignment)

The initial step in constructing a phylogenetic tree is the collection and assembly of nucleotide or protein sequence data (Lio and Goldman 1998; Linder and Warnow 2001). A variety of computational tools, many of which are open source, are available online to facilitate efficient analysis based on the research topic, data characteristics, and bioinformatics objectives. For example, tools like the National Center for Biotechnology Information (NCBI)'s Basic Local Alignment Search Tool (BLAST) enable the retrieval of DNA sequences by comparing them to query sequences. The extracted sequences are then evaluated based on their scores and E-values. A high score indicates a close match to the query sequence, while a low E-value suggests a high probability that the similarity represents a true evolutionary relationship rather than a chance occurrence. Typically, sequences with E-values of less than 10^{-5} are considered homologous to the query sequence (Altschul et al. 1997; Hall 2004).

After selecting and extracting the sequences, a multiple sequence alignment (MSA) is generated to identify conserved positions by arranging sequences within a matrix (Phillips et al. 2000; Edgar and Batzoglou 2006). Gaps, representing insertions or deletions over time, are introduced into the aligned sequences to reflect evolutionary changes (Corpet 1988; Phillips et al. 2000). The effectiveness of the MSA relies heavily on gap analysis, which determines where and how gaps are inserted (Corpet 1988; Simmons and Ochoterena 2000). Various web-based tools and software programs, such as ClustalW (Thompson et al. 1994), MUSCLE (Edgar 2004), MAFFT (Katoh et al. 2002), and T-Coffee (Notredame et al. 2000), facilitate MSAs of molecular data (Linder and Warnow 2001; Hall 2004).

3.1.2.2 Phylogenetic Tree Reconstruction Using Computational Methods and Probabilistic Models

Phylogenetic reconstruction methods are broadly classified into two categories: distance-based and character-based. Distance-based methods use substitution models for amino acids or nucleotides to create a distance matrix that reflects the similarities between sequences. This matrix forms the foundation for clustering and grouping sequences to construct phylogenetic trees. In contrast, character-based methods, such as ML, MP, and BI, evaluate the suitability and distribution of amino acid or nucleotide changes in phylogenies based on character substitution models. These methods often employ heuristic approaches to exclude suboptimal trees (Horner and Pesole 2004).

3.1.2.3 Statistical Evaluation and Validation of Inferred Trees

In recent decades, significant attention has been given to developing statistical validation methods for more accurate phylogenetic tree estimation. The resulting advancements have enabled researchers to determine which model best fits a given dataset. Statistical validation in phylogenetics assesses the likelihood that the topology of an inferred tree accurately represents true evolutionary relationships, thereby enhancing confidence in phylogenetic reconstructions and producing more realistic evolutionary models (Whelan et al. 2001). Phylogenetic inference algorithms typically generate multiple optimal trees, and statistical validation methods such as bootstrapping and jackknife resampling are used to identify the tree that most reliably represents evolutionary history (Felsenstein 2004).

3.2 Phylogenetic Tree Construction Methods

Phylogenetic trees are constructed using character datasets that shed light on evolutionary relationships. These datasets may include DNA, RNA, amino acids, molecular markers (such as single nucleotide polymorphisms (SNPs) and restriction fragment length polymorphisms (RFLPs)), and morphological data. Evolutionary modeling involves modifying nucleotide bases at specific positions within DNA sequences, where each character represents different states across taxa (Linder and Warnow 2001; Baldauf 2003; Dowell 2008). Phylogenetic tree construction primarily aims to delineate evolutionary relationships among organisms based on sequence data, estimating the time since divergence from a common ancestor shared by the species under study (Harrison and Langdale 2006). Achieving this goal relies on several fundamental assumptions: the accuracy of the sequences, their derivation from a common ancestral sequence, homology among aligned sites, and the presence of sufficient evolutionary signals within the sequence variations of the sample (Harrison and Langdale 2006). These foundational conditions are essential for accurate evolutionary analyses through various phylogenetic tree construction methods.

3.2.1 Distance-Based Methods

Distance-based methods construct phylogenetic trees by transforming sequence data into pairwise distances, which are then used to create a distance matrix. This matrix serves as the basis for tree construction. Notable examples of distance-based methods include the unweighted pair group method with arithmetic mean (UPGMA) for rooted trees and NJ for unrooted trees (Prager and Wilson 1978; Felsenstein 1984; Tamura et al. 2004).

One key advantage of distance-based methods, particularly NJ, is their computational efficiency. Unlike the MP and ML methods, which involve comparing numerous trees under optimal criteria, clustering algorithms such as NJ avoid exhaustive tree comparisons, producing results faster. Thus, NJ is especially effective for analyzing large datasets with relatively low sequence divergence.

However, distance-based methods, including NJ, can be susceptible to significant sampling errors when dealing with sequences that exhibit considerable evolutionary divergence. They may also show high variance in estimates due to large distance calculations and are sensitive to gaps within aligned sequences. Therefore, these factors must be considered carefully to obtain accurate results (Bruno et al. 2000; Tamura et al. 2004).

3.2.1.1 Unweighted Pair Group Method with Arithmetic Mean

UPGMA constructs a distance matrix from sequence data to represent differences between species. It builds phylogenetic trees by iteratively grouping the closest sequences and forming internal nodes until all sequences are connected. This method assumes that evolutionary rates are uniform across all branches, which implies similarity between species rather than providing precise evolutionary distances (Prager and Wilson 1978; Felsenstein 1988).

3.2.1.2 Neighbor-Joining

The NJ method aims to find a tree that minimizes the total branch length while computing distances between sequences. This widely used distance-based method seeks to construct an optimal tree where shorter evolutionary distances are reflected in better tree structures. Unlike UPGMA, NJ calculates internal node lengths directly to minimize the overall branch length without forming clusters. It starts with a star-shaped, unrooted tree with a single internal node, progressively connecting the most similar taxa. Branches are iteratively inserted between these taxa and the remaining ones to minimize the total branch length of the grouped neighbors. At each step, pairs of neighbors that reduce branch lengths the most are grouped to form clusters, ensuring a minimal sum of distances and branch lengths. This method assumes that sequences evolve at different rates, making it suitable for datasets with significant differences in evolutionary rates (Tamura et al. 2004; Yang and Rannala 2012). NJ is particularly effective for rapidly inferring phylogenetic trees from relatively clear datasets with low divergence. However, when inferring ancient evolutionary relationships, NJ-derived distance matrices may lack reliability, necessitating the consideration of alternative methods (Holder and Lewis 2003).

3.2.2 Character-Based Methods

Character-based methods estimate phylogenetic trees using aligned sequences, such as DNA or protein sequences, at each site. These methods assume that substitutions between sites are independent of adjacent sites, allowing each site to evolve independently. In these methods, homologous sites at corresponding positions in aligned sequences are used to infer evolutionary relationships from a common ancestor. Unlike distance-based methods, character-based approaches consider substitution probabilities between sequences, providing a detailed explanation of evolutionary processes. Prominent examples of character-based methods include MP and ML (Felsenstein 1981; Felsenstein 2004; Barton et al. 2007).

3.2.2.1 Maximum Parsimony

Originally developed for analyzing morphological traits, the parsimony method was adapted to molecular data in the 1970s. The Fitch-Hartigan algorithm, which reconstructs ancestral states by finding the minimum number of changes on a binary tree, is central to this method (Felsenstein 2004). MP operates on the principle that the phylogenetic tree with the fewest evolutionary changes best explains the observed data. However, as the number of sequences increases, the computational complexity grows exponentially due to the vast number of possible trees. Each alignment position contributes to all possible trees, with the most parsimonious tree being the one with the fewest evolutionary changes (Jin and Nei 1990; Takezaki and Nei 1994).

Parsimony does not utilize information from all sites but focuses on "informative sites"—those appearing at least twice with at least two different characters—affecting the tree's topology. These informative sites help construct the "parsimonious tree" by accumulating the minimum number of changes (Zhang and Nei 1997; Yang and Rannala 2012). Key programs for MP analysis include PAUP (Swofford 2000), MEGA (Tamura et al. 2011), and TNT (Goloboff et al. 2008).

While parsimony is relatively straightforward and less mathematically intensive than other methods, it lacks explicit assumptions about sequence evolution, complicating the integration of evolutionary knowledge into tree construction (Stewart 1993; Yang and Rannala 2012). Additionally, "long-branch attraction"—a systematic error where parsimony infers incorrect trees due to one short internal branch separating two long external branches—can lead to difficulties in accurately inferring multiple substitutions at the same sequence positions (Felsenstein 1978; Huelsenbeck 1997). To mitigate such issues, using more realistic substitution models or switching to ML methods may improve accuracy. Dense taxon sampling or excluding highly variable proteins or sequence positions can also help avoid generating long branches (Felsenstein 1978; Huelsenbeck 1997).

3.2.2.2 Maximum Likelihood

The ML method, developed by Ronald Aylmer Fisher in the 1920s, estimates model parameters based on probability (Fisher 1925). Joe Felsenstein developed the first ML algorithm for DNA sequence data in the 1980s (Felsenstein 1981). ML methods have since gained popularity due to advances in computational techniques and the development of realistic models of sequence evolution.

ML-based phylogenetic inference identifies the tree that best explains the variation in the dataset through probability calculations. It selects the tree with the highest likelihood among all possible trees generated from the data (Huelsenbeck and Crandall 1997). ML methods use various substitution models to estimate the likelihood of different phylogenies, assuming independent evolution at each sequence position (Felsenstein 2004; Yang 2006). While they generate all possible trees like MP methods, ML methods use "likelihood averages" across all possible states for each site, aiming to find the tree with the highest overall probability rather than the shortest evolutionary distances (Yang 1993, 1994).

ML methods provide more realistic tree estimates by considering variations in substitution rates rather than assuming a uniform "transition-transversion ratio" across all branches (Huelsenbeck and Crandall 1997; Pagel 1999). They are flexible in evaluating and refining evolutionary models, making them highly suitable for analyzing conserved protein sequences (Yang 1993, 1994; Blanquart and Lartillot 2008). Likelihood ratio tests can also be used to assess the suitability of evolutionary models or test biological hypotheses, such as molecular clock theory and Darwinian selection (Felsenstein 1981; Goldman 1993). However, ML methods depend on the chosen evolutionary model, and their accuracy may be compromised if the model is uncertain (Huelsenbeck and Rannala 1997; Whelan and Goldman 2001). Additionally, analyzing each position in multiple alignments can be complex and computationally intensive, requiring extensive computational resources. Phylogenetic software tools like PHYLIP (Felsenstein 1989), PAUP* (Swofford 2000), PhyML (Guindon et al. 2005), RAxML (Stamatakis 2014), and MEGA (Tamura et al. 2011) facilitate ML analysis.

3.2.3 DNA and Protein Evolution Models

The first DNA evolution model, known as the JC69 model, was proposed by Jukes and Cantor in 1969. This one-parameter model assumes equal frequency and substitution rates for all nucleotides. It postulates that substitutions occur randomly, with no insertions or deletions, and the only parameter is the overall substitution rate, normalized to a mean rate of 1 (Jukes and Cantor 1969). In 1980, Kimura introduced the K2P model, which includes two parameters: transition rates (purine to purine or pyrimidine to pyrimidine) and transversion rates (purine to pyrimidine or vice versa). Unlike JC69, the K2P model accommodates differences in substitution rates. Initially, Kimura used α and β to represent these rates, but it is now more common to set the transversion rate to 1 and the transition rate to k. This model assumes equal nucleotide frequencies ($\pi T = \pi C = \pi A = \pi G = 0.25$) (Kimura 1980).

In 1981, Felsenstein proposed the F81 model, which includes four parameters representing the substitution rates for each nucleotide (Felsenstein 1981). Subsequently, Hasegawa, Kishino, and Yano combined the K2P and F81 models to create the HKY85 model, which has five parameters (Hasegawa et al. 1985). The Tajima and Nei distance model allows for different substitution rates among nucleotides (Nei et al. 1983; Tajima and Nei 1984), while the gamma distance model assumes that nucleotide substitution rates follow a gamma distribution, specified by the parameter α, which is the inverse of the variance of substitution rates (Felsenstein 2004). The general time reversible (GTR) model assigns six different substitution rates, incorporating an equilibrium base frequency vector $\pi = (\pi T, \pi C, \pi A, \pi G)$ to represent nucleotide frequencies and a rate matrix for each position (Tavaré 1986).

For protein evolution models, the point accepted mutation (PAM) and blocks substitution matrix (BLOSUM) are commonly used (Henikoff and Henikoff 1992, 2000). PAM matrices are based on observed amino acid substitutions in closely related protein sequences and represent the probability of one amino acid substituting another over specific evolutionary periods. Different PAM matrices correspond to different evolutionary timescales. In contrast, BLOSUM matrices are based on local alignments of more distantly related protein sequences. These matrices are constructed from blocks of conserved sequences, with each number indicating the percentage of homology used to create the blocks. Typically, low PAM and high BLOSUM scores are used for recently diverged or longer query sequences, while high PAM and low BLOSUM scores are preferred for short or more divergent sequences. The number following the term "BLOSUM" denotes the percentage of homology used for block construction. Using a high BLOSUM score for short sequences may result in missing homologous blocks and false negatives, while PAM matrices generally perform better with shorter sequences. Although these guidelines may not always apply perfectly, protein substitution models can be selected based on these principles, with iterative testing of different matrices to optimize results (Henikoff and Henikoff 1992, 2000; Mount 2008a, b).

3.2.4 Statistical Validation of Phylogenetic Trees

Statistical validation methods such as jackknifing and bootstrapping are commonly employed to evaluate the accuracy and reliability of inferred phylogenetic trees. Jackknifing, specifically delete-half jackknifing, is a resampling technique that recalculates subsets of the original data (Farris et al. 1996). This method involves repeatedly removing a random half of the sequence sites from the original dataset to create new sequences of half the original length, which are then used to construct phylogenetic trees, referred to as subtrees. The frequency of these subtrees is calculated and expressed as a percentage or numerical value known as the jackknifing value. If each subset produces the same phylogenetic tree, its reliability is confirmed; if different trees are produced, reliability is diminished. A subtree appearing in all constructed trees would have a jackknifing value of 100%. Typically, branches with jackknifing values below 70% are considered to have questionable reliability (Felsenstein 1985; Mehta et al. 1995; Farris et al. 1996).

Bootstrapping is a computer-based technique used to evaluate the accuracy of statistical estimates. It was introduced by Felsenstein in 1985 to estimate phylogenetic trees (Felsenstein 1985). It measures the confidence of each clade in the observed tree based on the proportion of trees that produce the same clade. Bootstrapping involves resampling the alignment sequences to create new alignments of length n and repeatedly comparing the phylogenies of these samples. The consistency of branching patterns is quantified, and each internal node of the tree is assigned a bootstrap value. For example, if a bootstrap value of 100 is set, 100 sample trees are generated from the initial matrix, and a node with a bootstrap value of 90 indicates that the same topology was produced 90 out of 100 times. Bootstrap values between 90 and 100 are considered statistically significant, and branches with a bootstrap value of 95% or higher are generally deemed reliable in terms of topology (Felsenstein 1985; Hedges 1992; Hillis and Bull 1993; Soltis and Soltis 2003).

3.3 Applications

Computational methods for inferring and analyzing phylogenetic trees are essential in various applied biological fields, such as multiple sequence alignment, protein structure prediction, gene and protein function prediction, drug design, and drug target research. The scope and scale of obtaining new molecular phylogenetic information and using it in academic and industrial applications are expanding across numerous disciplines, including microbiology, infectious disease epidemiology, pharmacology, cancer genomics, forensic science, botany, marine biology, animal husbandry, and environmental science (Stamatakis 2005).

Phylogenetic methods have become highly valuable in molecular epidemiological research. Phylogenetic tree construction, alongside molecular clock analysis, is a crucial scientific tool for obtaining important information about major infectious diseases. For example, phylogenetic analysis has been used to support the claim that human immunodeficiency virus (HIV) has been present in human populations since the 1930s rather than spreading through contaminated polio vaccines (Worobey et al. 2004). Molecular phylogenetic methods have been employed to classify HIV-1 subtypes and estimate the divergence time of the common ancestor of the HIV-1 strain responsible for the acquired immune deficiency syndrome (AIDS) pandemic using ML (Kober et al. 2000; Horner and Pesole 2004).

Molecular phylogenetic analysis is also used in disease-related research involving human genomic information. Haplotype association studies are a prominent example of such research. Haplotypes are genotypes presumed to have been inherited intact as blocks of genetic sequences from ancient ancestors through subsequent generations. The association between haplotypes and disease susceptibility loci has been used to identify disease-related genes and loci. Phylogenetic trees explaining the vertical transmission of specific haplotypes provide branching information that can be explored to determine their association with disease susceptibility. Thus, molecular phylogenetic research methods enhance the efficiency and accuracy of identifying disease-related loci through their integrative use in genomics epidemiology (Horner and Pesole 2004).

In fields like botany, marine biology, animal husbandry, and wildlife conservation, phylogenetic studies enable the accurate classification of plants and animals, improve the quality of key agricultural and livestock products, and identify species or subspecies with conservation value to prevent hybridization. As the importance of securing biological resources grows globally, there is an increasing need to extensively collect, efficiently classify, and select these resources domestically. Consequently, using nucleotide sequence information to classify genetic resources and molecular markers related to valuable traits to select and classify useful ones is gaining prominence, with molecular phylogenetic analysis being essential for these tasks. As genomic information on biological resources continues to accumulate, the application of molecular phylogenetic analysis in evolutionary and taxonomic research is expected to become more widespread (Cho et al. 2008).

Recently, molecular phylogenetic methods have been applied to cancer research and therapy (Abu-Asab et al. 2008). Cancer development and progression can be considered evolutionary processes involving genetic modifications, selective pressure, and clonal propagation at the cellular and population levels (Sarnat and Netsky 1984; Azzone 1996; Abu-Asab et al. 2006). Therefore, evolutionary analysis methods hold significant potential as analytical tools in cancer diagnosis and research (Desper 2004; Abu-Asab et al. 2006; Pennington 2006). Specifically, molecular phylogenetic analysis enables the hierarchical classification of biological samples showing common mutations and pathway alterations related to cancer, thus identifying new gene and protein expression states that could serve as diagnostic biomarkers. Additionally, by elucidating the direction of changes driving molecular and cellular diversity among biological samples, these methods can provide insights into previously unidentified mechanisms of cancer development, facilitating new drug development (Abu-Asab et al. 2008).

3.4 Host Phylogeny and Virus Adaptation: Implications for Emerging Infectious Diseases

New viral variants, major causes of emerging or re-emerging infectious diseases, are often transmitted from their original native host species to humans, resulting in zoonotic infections. Notable examples include SARS-CoV, HIV, the swine-derived new influenza virus, Ebola virus, and MERS-CoV. These viral pathogens have caused large-scale infections in the human population due to changes in their host range (Woolhouse and Gowtage-Sequeria 2005). The emergence of new viral diseases in recent decades highlights the process of virus adaptation, which is closely related to the acquisition of altered cell tropism or host range (Pulliam 2008; Mänz et al. 2013). Therefore, it is crucial to identify pathogen, host, and environmental factors that contribute to viruses altering their host range to facilitate cross-species infections (Cho 2019).

The interrelationships between potential host species significantly influence the infectivity of new viruses crossing species barriers. Studies have shown that virus spillover is more likely when the taxonomic relationship between host species is close (Power and Mitchell 2004; Parrish et al. 2008; Plowright et al. 2015; Johnson et al. 2015). This insight has led to attempts to assess and predict viral infectivity in humans based on the interrelationships between host organisms (Lalić et al. 2011).

In the last decade, the identification of genome sequences for many host organisms and the development of new tools have enabled the study of virus-host interactions at the molecular level. Advances in high-throughput sequencing technology and bioinformatics methodologies have facilitated the analysis of gene or protein interaction networks that regulate host responses to pathogens. For instance, transcriptional profiling of simian immunodeficiency virus (SIV) infections in various primate host species revealed distinctions in the responses of natural reservoir hosts, which do not develop AIDS, and non-natural hosts, such as rhesus macaques and pig-tailed macaques, which develop AIDS like HIV-infected humans. These findings suggest that understanding host factors restricting HIV infection could help develop effective intervention strategies (Haagmans et al. 2009).

Host phylogeny is a crucial determinant of viral replication and persistence in new host systems, impacting the emergence of new diseases. Studies have shown that rabies viruses exhibit higher rates of cross-species transmission between closely related host species in the wild (Streicker et al. 2010), and primate lentivirus phylogenies imply preferred host switching between closely related organisms (Charleston and Robertson 2002). On the other hand, conserved host receptors may enable certain viruses to infect distantly related hosts (Baranowski et al. 2001; Woolhouse 2002). These findings indicate that host phylogenetic relationships account for significant differences in viral replication and persistence among host species. Additionally, host phylogeny has a strong effect independent of the distance from the original host, suggesting its role in predicting sources of newly emerging diseases. However, the phylogeny effect is more complex than merely causing most host-switching events between closely related host species (Longdon et al. 2011). Understanding the complex interactions between host phylogeny and virus adaptation is essential for identifying risk factors of emerging infectious diseases and developing strategies to mitigate their impact. This knowledge can enhance our ability to predict and manage the spread of new zoonotic infections, ultimately protecting public health (Cho 2019).

The following case study illustrates how phylogenetic analysis can be applied to elucidate the evolutionary relationships among viruses (Cho et al. 2022). This section focuses on the HIV-1, HIV-2, and SIV, analyzing the *gag*, *pol*, and *env* genes, which are common structural genes of lentiviruses. For additional case studies involving the phylogenetic analysis of different viruses, please refer to "Part III. Applications and Case Studies".

3.5 Case Study: Evolutionary Relationships Among Primate Lentiviruses

HIV belongs to the genus *Lentivirus*, which is part of the *Orthoretrovirus* subfamily within the *Retrovirus* family and currently includes a total of 10 species (Coffin et al. 2021). Among these, HIV-1 and HIV-2, which infect humans and cause AIDS, along with SIV, which infects non-human primates, comprise the primate lentivirus group. Members of the genus *Lentivirus* universally contain the *gag*, *pol*, and *env* genes, although the number and types of accessory proteins vary among species. The *gag* gene encodes the matrix (MA), capsid (CA), and nucleocapsid (NC) proteins, the *env* gene encodes the surface protein (SU) and transmembrane protein (TM), and the *pol* gene encodes three essential enzymes for the viral lifecycle: reverse transcriptase, integrase, and protease. The extent of variation differs among these genes, with *gag* and *pol* being more conserved than *env* (Foley 2000). Phylogenetic trees constructed for the *gag*, *pol*, and *env* genes have confirmed that the primate lentivirus group forms a distinct clade (Sharp et al. 1999). A key difference between primate lentiviruses and other lentiviruses is the absence of a dUTPase-coding region within the *pol* gene in the former (Foley 2000). The dUTPase enzyme decomposes dUTP to reduce the dUTP/dTTP ratio in cells, thus minimizing misincorporation during DNA replication and ensuring accuracy (Hizi and Herzig 2015). Instead of dUTPase, primate lentiviruses use alternative mechanisms to ensure viral replication. For instance, while uracilization during reverse transcription in HIV is high, it prevents autointegration and promotes the insertion of viral DNA into the host cell chromosome, facilitating viral replication (Yan et al. 2011). These unique characteristics of primate lentiviruses, including HIV, enhance their infectivity in human hosts.

HIV is classified into two genetically distinct types: HIV-1 and HIV-2. While they share transmission routes and pathological processes, HIV-2 exhibits lower pathogenicity and transmissibility, a longer latency period, a lower likelihood of progressing to AIDS, and lower viral loads than HIV-1 (Vidyavijayan et al. 2017). Unlike the spread of HIV-1 to pandemic levels, HIV-2 is primarily concentrated in West Africa but has also been reported in countries like Angola, Mozambique, Portugal, France, the United States, and India (Visseaux et al. 2016). Sequence analysis shows that HIV-1 and HIV-2 share only up to 55% homology (Motomura et al. 2008) and encode different accessory proteins. In contrast, SIV has been identified in over 40 non-human primate species. In natural hosts like the sooty mangabey, SIV can achieve high viral loads like HIV infection without causing disease (Rey-Cuillé et al. 1998). However, when transmitted across species, such as to chimpanzees or Asian macaques (*Macaca mulatta*), SIV can cause simian AIDS (Apetrei et al. 2005). Non-pathogenic SIV strains do not cause significant disease in their natural hosts, likely due to a weaker immune response and their use of only the CCR5 coreceptor, as opposed to the CXCR4 coreceptor utilized by pathogenic strains (Holmes 2001). These differences highlight the genetic and epidemiological differences between HIV-1 and HIV-2 and the varying, host-dependent infection patterns of SIV.

Previous phylogenetic studies on the *gag*, *pol*, and *env* genes of primate lentiviruses have shown that rather than forming two large clades based on the host (human and non-human primates), HIV-1, SIVcpz, and SIVgor form one clade, while HIV-2 and SIVsm form another. Additionally, HIV-2 has been observed to be concentrated in West Africa, where sooty mangabeys reside and are kept as pets in rural areas. This suggests a high probability that HIV-2 was transmitted to humans from SIVsm crossing the species barrier (Sharp et al. 1999). The phylogenetic tree also shows that within the HIV-2 and SIVsm clade, HIV-2 lineages are not clustered together but intermixed with SIVsm. This indicates that HIV-2 was transmitted to humans from sooty mangabeys through at least eight separate cross-species transmission events. Similarly, epidemiological and phylogenetic analyses have confirmed that the M and N groups of HIV-1 were transmitted to humans through at least two cross-species events from SIVcpz (Gao et al. 1999), while the O and P groups were transmitted from SIVgor (Sharp and Hahn 2010; Hemelaar 2012). Old World monkeys are known to be the natural hosts of SIV. SIVcpz, found in chimpanzees, is a recombinant of SIVrcm and SIVgsn (Sharp et al. 2005), and SIVgor was transmitted to gorillas from chimpanzees (Takehisa et al. 2009). Initially, SIVcpz was thought not to cause disease in chimpanzees, but a study by Keele et al. in 2009 revealed that, like HIV-1, SIVcpz leads to progressive CD4+ T cell loss, early death, and negative effects on health and reproduction in chimpanzees. Chimpanzees and Asian macaques that acquired SIV through cross-species transmission from Old World monkeys exhibit AIDS-like symptoms similar to those seen in humans (Neel et al. 2010). Consequently, macaques are widely used as animal models in HIV research (Gardner and Luciw 2008).

The following example of phylogenetic analysis confirms the evolutionary relationships between primate lentiviruses isolated from human hosts (HIV-1 and HIV-2) and various non-human primate hosts (SIV) (Cho et al. 2022). The taxonomies of SIV hosts and SIV variants according to host species are shown in Table 3.1. Considering the group SIVcer, with the minimum number of available sequences (n = 5), phylogenetic trees were repeatedly created by building datasets with sequences randomly extracted from the sequence pool for each group. The clustering pattern was confirmed to be consistent in all trees reconstructed in this way. Regarding SIV, some host species only had one sequence available; therefore, 60, 59, and 53 sequences for each *gag*, *pol*, and *env* gene, respectively, were used to construct the final phylogenetic tree. After MSA using ClustalW (Thompson et al. 1994), phylo-

Table 3.1 Taxonomic classification of SIV hosts

Subfamily	Tribe	Genus	Species	SIV abbreviation
Cercopithecinae	Cercopithecini	Allochrocebus	lhoesti	SIVlst
			solatus	SIVsun
		Chlorocebus	aethiops	SIVagm-grv
			cynosuros	SIVagm-mal
			sabaeus	SIVagm-sab
			tantalus	SIVagm-tan
		Cercopithecus	albogularis	SIVsyk
			ascanius	SIVasc
			ascanius_whitesidei	SIVasw
			cephus	SIVmus
			mitis	SIVblu
			mona	SIVmon
			neglectus	SIVdeb
			nictitans	SIVgsn
		Miopithecus	ogouensis	SIVtal
	Papioini	Cercocebus	agilis	SIVagi
			atys	SIVsmm
			torquatus	SIVrcm
		Macaca	arctoides	SIVstm
			fascicularis	SIVcem
			mulatta	SIVmac
			nemestrina	SIVmne
Cercopithecinae	Papioini	Mandrillus	sphinx	SIVmnd-1
				SIVmnd-2
			leucophaeus	SIVdrl
Colobinae		Colobus	guereza	SIVcol
		Piliocolobus	badius_badius	SIVwrc
			badius_temminckii	SIVtrc
		Procolobus	verus	SIVolc
Homininae	Gorillini	Gorilla	gorilla gorilla	SIVgor
	Hominini	Pan	troglodytes_schweinfurthii	SIVcpz-eas
			troglodytes_troglodytes	SIVcpz-cen
			troglodytes_verus	SIVcpz-wes

Species are represented by their Latin names, with the genus name preceding the standard name of the species

genetic trees were constructed using the ML method (Kishino and Hasegawa 1989) based on the Kimura-2 parameter model through MEGA-X (Kumar et al. 2018). The bootstrap parameter was set to 1000 to measure the reliability of each internal node. To minimize possible bias among the phylogenetic trees, this process was repeated five times for each *gag*, *pol*, and *env* gene, resulting in a representative tree with a high degree of agreement.

Phylogenetic analyses of *gag*, *pol*, and *env* confirmed that HIV-1 and HIV-2 clustered with the SIV species identified as their respective origins (Fig. 3.3). Separate phylogenetic analyses of each gene showed that HIV-1 formed clades with SIVcpz and SIVgor, while HIV-2 formed clades with SIVsmm and SIVmacaca. SIVcpz has been proposed to be a recombinant of SIVgsn (3′) and SIVrcm (5′) (Sharp et al. 2005), and the gene tree constructed in this study supported this assumption, showing that SIVcpz clustered close to SIVgsn and SIVrcm in the *env* and *pol* gene trees. Furthermore, SIVlst and SIVsun, whose hosts belong to the genus *Allochrocebus*, were closely

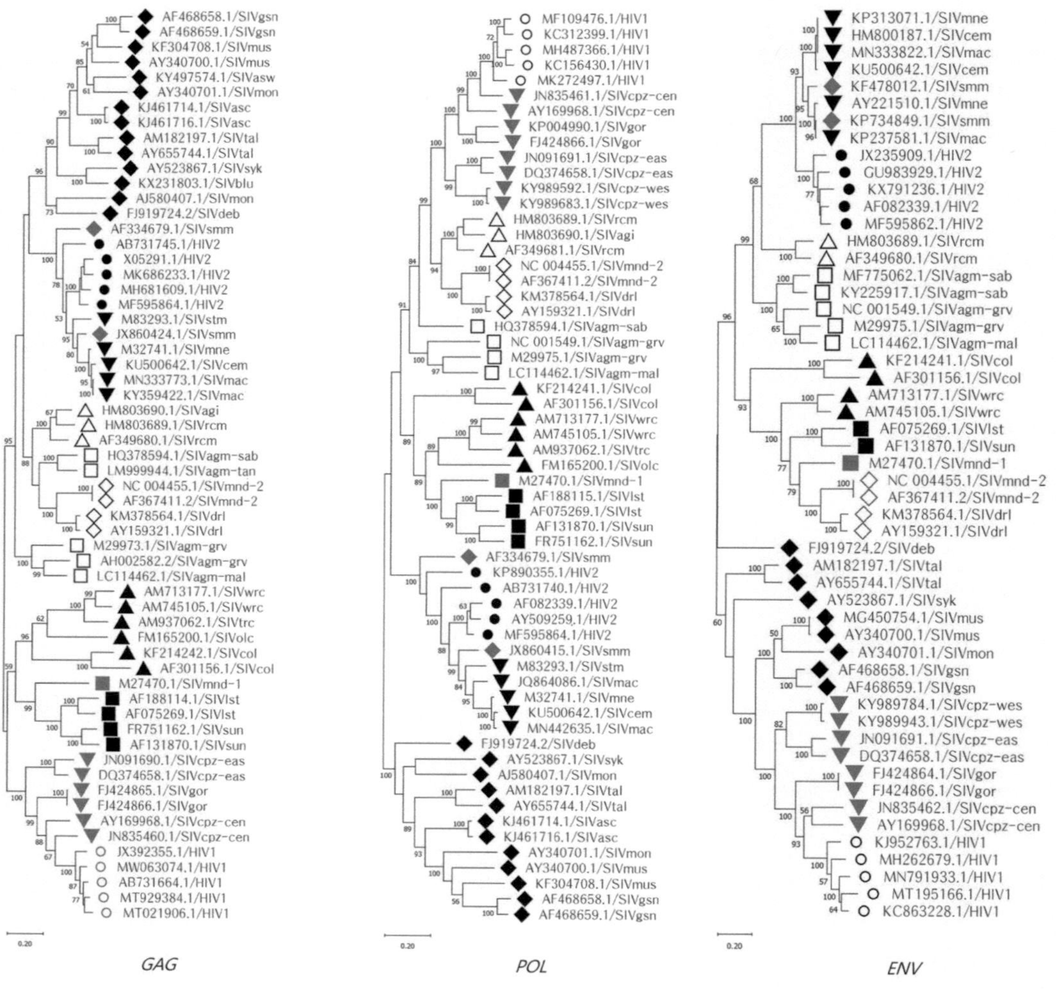

Fig. 3.3 Phylogenetic trees based on nucleotide sequences of lentivirus structural genes. Maximum likelihood trees show evolutionary relationships among primate lentiviruses based on the nucleotide sequences of lentivirus structural genes (*gag*, *pol*, and *env*). Bootstrap values >70% (1000 replications) are shown at the nodes, and branch lengths indicate evolutionary distances. Symbols represent virus species: [○] HIV-1; [●] HIV-2; [□] SIVagm; [■] SIVallo; [△] SIVcer; [▲] SIVcol; [▼] SIVcpz_gor; [▼] SIVmac; [◇]SIVmnd; [◆] SIVpit; [◆] SIVsmm

related to SIVcolobus. Based on the three phylogenetic trees, the *env* gene exhibited greater variation than the *gag* and *pol* genes. Given that the *env* gene encodes the HIV envelope glycoprotein, a target for cytotoxic T lymphocytes and neutralizing antibodies, the distinctive level and pattern of *env* gene variations compared to the *gag* and *pol* genes may be due to selection for antigenic diversity. Selective pressure on the *env* gene exerted by cell-mediated immunity results in escape mutations. Thus, the host's antibody response can continuously drive viral evolution with the initial response to the transmitted virus and sequential responses to escape variants (Richman et al. 2003). Additionally, the *env* gene can evolve to encode envelope glycoproteins for interaction with altered receptors or coreceptors, allowing the virus to enter alternative host cells, further supporting the distinctive level and pattern of *env* gene variations compared to other genes. In most instances where the host monkey species were closely related (i.e., they belonged to the same genus), the corresponding SIV species were also closely related.

3.6 Conclusion

With the growing availability of molecular sequence data, methodological advancements in phylogenetics, and the development of various evolutionary models, highly accurate and reliable evolutionary analyses are now possible. These advancements not only promise valuable outcomes but also broaden the scope of applications. To derive the best evolutionary inferences using the diverse analytical methods currently available, it is essential to systematically perform dataset construction and precise MSA and select the optimal analysis methods. The rapid accumulation of vast genetic datasets, fuelled by advances in genome sequencing technology, has made molecular phylogenetics a key component of many biological studies. Genome-scale studies on genetic information, conserved gene order, gene expression, regulatory networks, metabolic pathways, and functional genome annotation can be significantly enhanced by phylogenetic analysis-based evolutionary research. Molecular phylogenetics has become an integral part of bioinformatics approaches for various biological research areas, drug design, gene characterization, and protein structure prediction. Molecular phylogenetic analyses are actively contributing to the identification of infectious microorganisms, the development of vaccines, the prediction of the timing and types of emerging infectious disease epidemics, the discovery of molecular markers or useful substances for diagnosis and treatment, drug targeting, and the development of new therapeutics. The integration of genomics, evolutionary analysis, and medical science can provide new perspectives and insights for addressing potential future public health challenges. Therefore, ongoing research and development efforts, including the design of new algorithms to solve the many complex issues in phylogenetic inference, must continue to keep pace with their rapidly expanding scope.

References

Abu-Asab M, Chaouchi M, Amri H. Phyloproteomics: what phylogenetic analysis reveals about serum proteomics. J Proteome Res. 2006;5:2236–40.

Abu-Asab M, Chaouchi M, Amri H. Evolutionary medicine: a meaningful connection between omics, disease, and treatment. Proteomics Clin Appl. 2008;2:122–34.

Altschul SF, Madden TL, Schäffer AA, Zhang J, Zhang Z, Miller W, Lipman DJ. Gapped BLAST and PSI-BLAST: a new generation of protein database search programs. Nucleic Acids Res. 1997;25:3389–402.

Apetrei C, Kaur A, Lerche NW, Metzger M, Pandrea I, Hardcastle J, Falkenstein S, Bohm R, Koehler J, Traina-Dorge V, Williams T, Staprans S, Plauche G, Veazey RS, McClure H, Lackner AA, Gormus B, Robertson DL, Marx PA. Molecular epidemiology of simian immunodeficiency virus SIVsm in US primate centers unravels the origin of SIVmac and SIVstm. J Virol. 2005;79:8991–9005.

Azzone GF. The nature of diseases: evolutionary, thermodynamic and historical aspects. Hist Philos Life Sci. 1996;18:83–106.

Baldauf SL. Phylogeny for the faint of heart: a tutorial. Trends Genet. 2003;19:345–51.

Baranowski E, Ruiz-Jarabo CM, Domingo E. Evolution of cell recognition by viruses. Science. 2001;292:1102–5.

Barton NH, Briggs DE, Eisen JA, Goldstein DB, Patel NH. Phylogenetic reconstruction. Evolution. 2007;61:1–55.

Blanquart S, Lartillot N. A site-and time-heterogeneous model of amino acid replacement. Mol Biol Evol. 2008;25:842–58.

Bruno WJ, Socci ND, Halpern AL. Weighted neighbor joining: a likelihood-based approach to distance-based phylogeny reconstruction. Mol Biol Evol. 2000;17:189–97.

Charleston MA, Robertson DL. Preferential host switching by primate lentiviruses can account for phylogenetic similarity with the primate phylogeny. Syst Biol. 2002;51:528–35.

Cho M. Prediction study on the cross-species infection of viruses using bioinformatics methods. PhD Thesis, Graduate School of Public Health, Seoul National University, Republic of Korea, 2019.

Cho M, Son HS. Molecular phylogenetics for bioinformatics. JCBB. 2017;2:90–100.

Cho SR, Seol SS, Park JM. An introduction of management and policy of biological resources. J Korea Technol Innov Soc. 2008;11:219–40.

Cho M, Min X, Son HS. Analysis of evolutionary and genetic patterns in structural genes of primate lentiviruses. Genes Genom. 2022;44:773–91.

Coffin J, Blomberg J, Fan H, Gifford R, Hatziioannou T, Lindemann D, Mayer J, Stoye J, Tristem M, Johnson W, ICTV Report Consortium. ICTV virus taxonomy profile: Retroviridae 2021. J Gen Virol. 2021;102:001712.

Corpet F. Multiple sequence alignment with hierarchical clustering. Nucleic Acids Res. 1988;16:10881–90.

Desper R, Khan J, Schäffer AA. Tumor classification using phylogenetic methods on expression data. J Theor Biol. 2004;228:477–96.

Dowell K. Molecular phylogenetics: An introduction to computational methods and tools for analyzing evolutionary relationships. Mol Phylogenetics. 2008;1–19.

Durbin R, Eddy SR, Krogh A, Mitchison G. Biological sequence analysis: probabilistic models of proteins and nucleic acids. Cambridge University Press. 1998.

Edgar RC. MUSCLE: multiple sequence alignment with high accuracy and high throughput. Nucleic Acids Res. 2004;32:1792–7.

Edgar RC, Batzoglou S. Multiple sequence alignment. Curr Opin Struct Biol. 2006;16:368–73.

Edwards SV. Is a new and general theory of molecular systematics emerging?. Evolution. 2009;63:1–19.

Farris JS, Albert VA, Källersjö M, Lipscomb D, Kluge AG. Parsimony jackknifing outperforms neighbor-joining. Cladistics. 1996;12:99–124.

Felsenstein J. Cases in which parsimony or compatibility methods will be positively misleading. Syst Zool. 1978;27:401–10.

Felsenstein J. Evolutionary trees from DNA sequences: a maximum likelihood approach. J Mol Evol. 1981;17:368–76.

Felsenstein J. Distance methods for inferring phylogenies: a justification. Evolution. 1984;16–24.

Felsenstein J. Confidence limits on phylogenies: an approach using the bootstrap. Evolution. 1985;39:783–91.

Felsenstein J. Phylogenies from molecular sequences: inference and reliability. Annu Rev Genet. 1988;22:521–65.

Felsenstein J. PHYLIP-phylogeny inference package (Ver. 3.2). Cladistics. 1989;5:164–6.

Felsenstein J. Inferring phylogenies. In: Inferring phylogenies. 2004;664.

Fisher RA. Theory of statistical estimation. Math Proc Camb Philos Soc. 1925;22:700–25.

Fitch WM, Margoliash E. Construction of phylogenetic trees. Science. 1967;155:279–84.

Foley BT. An overview of the molecular phylogeny of lentiviruses. In: HIV Sequence Compendium; 2000. p. 35–43.

Gao F, Bailes E, Robertson DL, Chen Y, Rodenburg CM, Michael SF, Cummins LB, Arthur LO, Peeters M, Shaw GM, Sharp PM, Hahn BH. Origin of HIV-1 in the chimpanzee *Pan troglodytes troglodytes*. Nature. 1999;397:436.

Gardner MB, Luciw PA. Macaque models of human infectious disease. ILAR J. 2008;49:220–55.

Goldman N. Statistical tests of models of DNA substitution. J Mol Evol. 1993;36:182–98.

Goloboff PA, Farris JS, Nixon KC. TNT, a free program for phylogenetic analysis. Cladistics. 2008;24:774–86.

Green RE, Krause J, Briggs AW, Maricic T, Stenzel U, Kircher M, Patterson N, Li H, Zhai W, Fritz MH-Y, Hansen NF, Durand EY, Malaspinas A-S, Jensen JD, Marques-Bonet T, Alkan C, Prüfer K, Meyer M, Burbano HA, Good JM, Schultz R, Aximu-Petri A, Butthof A, Höber B, Höffner B, Siegemund M, Weihmann A, Nusbaum C, Lander ES, Russ C, Novod N, Affourtit J, Egholm M, Verna C, Rudan P, Brajkovic D, Kucan Ž, Gušic I, Doronichev VB, Golovanova LV, Lalueza-Fox C, de la Rasilla M, Fortea J, Rosas A, Schmitz RW, Johnson PLF, Eichler EE, Falush D, Birney E, Mullikin JC, Slatkin M, Nielsen R, Kelso J, Lachmann M, Reich D, Pääbo S. A draft sequence of the Neandertal genome. Science. 2010;328:710–22.

Grenfell BT, Pybus OG, Gog JR, Wood JL, Daly JM, Mumford JA, Holmes EC. Unifying the epidemiological and evolutionary dynamics of pathogens. Science. 2004;303:327–32.

Gronau I, Hubisz MJ, Gulko B, Danko CG, Siepel A. Bayesian inference of ancient human demography from individual genome sequences. Nat Genet. 2011;43:1031–4.

Guindon S, Lethiec F, Duroux P, Gascuel O. PHYML online—a web server for fast maximum likelihood-based phylogenetic inference. Nucleic Acids Res. 2005;33(suppl 2):W557–9.

Haagmans BL, Andeweg AC, Osterhaus AD. The application of genomics to emerging zoonotic viral diseases. PLoS Pathog. 2009;5:e1000557.

Hall BG. Phylogenetic trees made easy: a how-to manual, vol. 547. Sunderland: Sinauer Associates; 2004.

Harrison CJ, Langdale JA. A step by step guide to phylogeny reconstruction. Plant J. 2006;45:561–72.

Hasegawa M, Kishino H, Yano TA. Dating of the human-ape splitting by a molecular clock of mitochondrial DNA. J Mol Evol. 1985;22:160–74.

Hedges SB. The number of replications needed for accurate estimation of the bootstrap P value in phylogenetic studies. Mol Biol Evol. 1992;9:366–9.

Hemelaar J. The origin and diversity of the HIV-1 pandemic. Trends Mol Med. 2012;18:182–92.

Henikoff S, Henikoff JG. Amino acid substitution matrices from protein blocks. Proc Natl Acad Sci USA. 1992;89:10915–9.

Henikoff S, Henikoff JG. Amino acid substitution matrices. Adv Protein Chem. 2000;54:73–98.

Hillis DM, Bull JJ. An empirical test of bootstrapping as a method for assessing confidence in phylogenetic analysis. Syst Biol. 1993;42:182–92.

Hillis DM, Huelsenbeck JP. Signal, noise, and reliability in molecular phylogenetic analyses. J Hered. 1992;83:189–95.

Hillis DM, Huelsenbeck JP, Cunningham CW. Application and accuracy of molecular phylogenies. Science. 1994;264:671–7.

Hizi A, Herzig E. dUTPase: the frequently overlooked enzyme encoded by many retroviruses. Retrovirology. 2015;12:1–15.

Holder M, Lewis PO. Phylogeny estimation: traditional and Bayesian approaches. Nat Rev Genet. 2003;4:275–84.

Holmes EC. On the origin and evolution of the human immunodeficiency virus (HIV). Biol Rev. 2001;76:239–54.

Horner DS, Pesole G. Phylogenetic analyses: a brief introduction to methods and their application. Expert Rev Mol Diagn. 2004;4:339–50.

Huelsenbeck JP. Is the Felsenstein zone a fly trap? Syst Biol. 1997;46:69–74.

Huelsenbeck JP, Crandall KA. Phylogeny estimation and hypothesis testing using maximum likelihood. Annu Rev Ecol Evol Syst. 1997;28:437–66.

Huelsenbeck JP, Rannala B. Phylogenetic methods come of age: testing hypotheses in an evolutionary context. Science. 1997;276:227–32.

Jin L, Nei M. Limitations of the evolutionary parsimony method of phylogenetic analysis. Mol Biol Evol. 1990;7:82–102.

Johnson CK, Hitchens PL, Evans TS, Goldstein T, Thomas K, Clements A, Joly DO, Wolfe ND, Daszak P, Karesh WB, Mazet JK. Spillover and pandemic properties of zoonotic viruses with high host plasticity. Sci Rep. 2015;5:14830.

Jukes TH, Cantor CR. Evolution of protein molecules. In: Munro HN, editor. Mammalian Protein Metabolism; 1969. p. 21–132.

Katoh K, Misawa K, Kuma KI, Miyata T. MAFFT: a novel method for rapid multiple sequence alignment based on fast Fourier transform. Nucleic Acids Res. 2002;30:3059–66.

Keele BF, Jones JH, Terio KA, Estes JD, Rudicell RS, Wilson ML, Li Y, Learn GH, Beasley TM, Schumacher-Stankey J, Wroblewski E, Mosser A, Raphael J, Kamenya S, Lonsdorf EV, Travis DA, Mlengeya T, Kinsel MJ, Else JG, Silvestri G, Goodall J, Sharp PM, Shaw GM, Pusey AE, Hahn BH. Increased mortality and AIDS-like immunopathology in wild chimpanzees infected with SIVcpz. Nature. 2009;460:515–9.

Kellis M, Patterson N, Endrizzi M, Birren B, Lander ES. Sequencing and comparison of yeast species to identify genes and regulatory elements. Nature. 2003;423:241–54.

Kimura M. A simple method for estimating evolutionary rates of base substitutions through comparative studies of nucleotide sequences. J Mol Evol. 1980;16:111–20.

Kishino H, Hasegawa M. Evaluation of the maximum likelihood estimate of the evolutionary tree topologies from DNA sequence data, and the branching order in Hominoidea. J Mol Evol. 1989;29:170–9.

Korber B, Muldoon M, Theiler J, Gao F, Gupta R, Lapedes A, Hahn BH, Wolinsky S, Bhattacharya T. Timing the ancestor of the HIV-1 pandemic strains. Science. 2000;288:1789–96.

Kumar S, Stecher G, Li M, Knyaz C, Tamura K. MEGA X: molecular evolutionary genetics analysis across computing platforms. Mol Biol Evol. 2018;35:1547–9.

Lalić J, Cuevas JM, Elena SF. Effect of host species on the distribution of mutational fitness effects for an RNA virus. PLoS Genet. 2011;7:e1002378.

Lemmon AR, Brown JM, Stanger-Hall K, Lemmon EM. The effect of ambiguous data on phylogenetic estimates obtained by maximum likelihood and Bayesian inference. Syst Biol. 2009;58:130–45.

Lewis PO. A genetic algorithm for maximum-likelihood phylogeny inference using nucleotide sequence data. Mol Biol Evol. 1998;15:277–83.

Li WH. Molecular Evolution. Sunderland, Mass: Sinauer Associates; 1997.

Li H, Durbin R. Inference of human population history from individual whole-genome sequences. Nature. 2011;475:493–6.

Lindblad-Toh K, Garber M, Zuk O, et al. A high-resolution map of human evolutionary constraint using 29 mammals. Nature. 2011;478:476–82.

Linder CR, Warnow T. An overview of phylogeny reconstruction, 2001.

Lio P, Goldman N. Models of molecular evolution and phylogeny. Genome Res. 1998;8:1233–44.

Longdon B, Hadfield JD, Webster CL, Obbard DJ, Jiggins FM. Host phylogeny determines viral persistence and replication in novel hosts. PLoS Pathog. 2011;7:e1002260.

Ma J. Reconstructing the history of large-scale genomic changes: biological questions and computational challenges. J Comput Biol. 2011;18:879–93.

Mänz B, Schwemmle M, Brunotte L. Adaptation of avian influenza A virus polymerase in mammals to overcome the host species barrier. J Virol. 2013;87:7200–9.

Marra MA, et al. The genome sequence of the SARS-associated coronavirus. Science. 2003;300:1399–404.

Mäser P, Thomine S, Schroeder JI, Ward JM, Hirschi K, Sze H, Talke IN, Amtmann A, Maathuis FJM, Sanders D, Harper JF, Tchieu J, Gribskov M, Persans

MW, Salt DE, Kim SA, Guerinot ML. Phylogenetic relationships within cation transporter families of Arabidopsis. Plant Physiol. 2001;126:1646–67.

Mehta PK, Heringa J, Argos P. A simple and fast approach to prediction of protein secondary structure from multiply aligned sequences with accuracy above 70%. Protein Sci. 1995;4:2517–25.

Motomura K, Chen J, Hu WS. Genetic recombination between human immunodeficiency virus type 1 (HIV-1) and HIV-2, two distinct human lentiviruses. J Virol. 2008;82:1923–33.

Mount DW. Using BLOSUM in sequence alignments. Cold Spring Harb Protoc. 2008a;2008:pdb-top39.

Mount DW. Using PAM matrices in sequence alignments. Cold Spring Harb Protoc. 2008b;2008:pdb-top38.

Neel C, Etienne L, Li Y, Takehisa J, Rudicell RS, Bass IN, Moudindo J, Mebenga A, Esteban A, Van Heuverswyn F, Liegeois F, Kranzusch PJ, Walsh PD, Sanz CM, Morgan DB, Ndjango JN, Plantier J, Locatelli S, Gonder MK, Leendertz FH, Boesch C, Todd A, Delaporte E, Mpoudi-Ngole E, Hahn BH, Peeters M. Molecular epidemiology of simian immunodeficiency virus infection in wild-living gorillas. J Virol. 2010;84:1464–76.

Nei M, Tajima F, Tateno Y. Accuracy of estimated phylogenetic trees from molecular data. J Mol Evol. 1983;19:153–70.

Notredame C, Higgins DG, Heringa J. T-Coffee: a novel method for fast and accurate multiple sequence alignment. J Mol Biol. 2000;302:205–17.

Pagel M. The maximum likelihood approach to reconstructing ancestral character states of discrete characters on phylogenies. Syst Biol. 1999;48:612–22.

Parrish CR, Holmes EC, Morens DM, Park EC, Burke DS, Calisher CH, Laughlin CA, Saif LJ, Daszak P. Cross-species virus transmission and the emergence of new epidemic diseases. Microbiol Mol Biol Rev. 2008;72:457–70.

Paten B, Herrero J, Fitzgerald S, Beal K, Flicek P, Holmes I, Birney E. Genome-wide nucleotide-level mammalian ancestor reconstruction. Genome Res. 2008;18:1829–43.

Pedersen JS, Bejerano G, Siepel A, Rosenbloom K, Lindblad-Toh K, Lander ES, Kent J, Miller W, Haussler D. Identification and classification of conserved RNA secondary structures in the human genome. PLoS Comput Biol. 2006;2:e33.

Pennington G, Smith CA, Shackney S, Schwartz R. Expectation-maximization method for reconstructing tumor phylogenies from single-cell data. In: Computational systems bioinformatics conference (CSB); 2006, July. p. 371–80.

Phillips A, Janies D, Wheeler W. Multiple sequence alignment in phylogenetic analysis. Mol Phylogenet Evol. 2000;16:317–30.

Plowright RK, Eby P, Hudson PJ, Smith IL, Westcott D, Bryden WL, Middleton D, Reid PA, McFarlane RA, Martin G, Tabor GM, Skerratt LF, Anderson DL, Crameri G, Quammen D, Jordan D, Freeman P, Wang LF, Epstein JH, Marsh GA, Kung NY, McCallum H. Ecological dynamics of emerging bat virus spillover. Proc Biol Sci. 2015;282(1798):20142124.

Power AG, Mitchell CE. Pathogen spillover in disease epidemics. Am Nat. 2004;164(S5):S79–89.

Prager EM, Wilson AC. Construction of phylogenetic trees for proteins and nucleic acids: empirical evaluation of alternative matrix methods. J Mol Evol. 1978;11:129–42.

Pulliam JRC. Viral host jumps: moving toward a predictive framework. EcoHealth. 2008;5:80–91.

Rey-Cuille MA, Berthier JL, Bomsel-Demontoy MC, Chaduc Y, Montagnier L, Hovanessian AG, Chakrabarti LA. Simian immunodeficiency virus replicates to high levels in sooty mangabeys without inducing disease. J Virol. 1998;72:3872–86.

Richman DD, Wrin T, Little SJ, Petropoulos CJ. Rapid evolution of the neutralizing antibody response to HIV type 1 infection. Proc Natl Acad Sci. 2003;100:4144–9.

Saitou N, Imanishi T. Relative efficiencies of the Fitch-Margoliash, maximum-parsimony, maximum-likelihood, minimum-evolution, and neighbor-joining methods of phylogenetic tree construction in obtaining the correct tree. Mol Biol Evol. 1989;6:514–25.

Saitou N, Nei M. The neighbor-joining method: a new method for reconstructing phylogenetic trees. Mol Biol Evol. 1987;4:406–25.

Sarnat HB, Netsky MG. Hypothesis: phylogenetic diseases of the nervous system. Can J Neurol Sci. 1984;11:29–33.

Sharp PM, Hahn BH. The evolution of HIV-1 and the origin of AIDS. Philos Trans R Soc Lond Ser B Biol Sci. 2010;365:2487–94.

Sharp PM, Bailes E, Robertson DL, Gao F, Hahn BH. Origins and evolution of AIDS viruses. Biol Bull. 1999;196:338–42.

Sharp PM, Shaw GM, Hahn BH. Simian immunodeficiency virus infection of chimpanzees. J Virol. 2005;79:3891–902.

Simmons MP, Ochoterena H. Gaps as characters in sequence-based phylogenetic analyses. Syst Biol. 2000;49:369–81.

Soltis PS, Soltis DE. Applying the bootstrap in phylogeny reconstruction. Stat Sci. 2003;18:256–67.

Stamatakis A. Phylogenetics: Applications, software and challenges. Cancer Genomics & Proteomics. 2005;2:301–5.

Stamatakis A. RAxML version 8: a tool for phylogenetic analysis and post-analysis of large phylogenies. Bioinformatics. 2014;30:1312–3.

Stewart CB. The powers and pitfalls of parsimony. Nature. 1993;361:603–7.

Streicker DG, Turmelle AS, Vonhof MJ, Kuzmin IV, McCracken GF, Rupprecht CE. Host phylogeny constrains cross-species emergence and establishment of rabies virus in bats. Science. 2010;329:676–9.

Swofford DL. *PAUP*: phylogenetic analysis by parsimony (and other methods) 4.0 Beta, 2000.

Tajima F, Nei M. Estimation of evolutionary distance between nucleotide sequences. Mol Biol Evol. 1984;1:269–85.

Takahashi K, Nei M. Efficiencies of fast algorithms of phylogenetic inference under the criteria of maximum parsimony, minimum evolution, and maximum likelihood when a large number of sequences are used. Mol Biol Evol. 2000;17:1251–8.

Takehisa J, Kraus MH, Ayouba A, Bailes E, Van Heuverswyn F, Decker JM, Li Y, Rudicell RS, Learn GH, Neel C, Ngole EM, Shaw GM, Peeters M, Sharp PM, Hahn BH. Origin and biology of simian immunodeficiency virus in wild-living western gorillas. J Virol. 2009;83:1635–48.

Takezaki N, Nei M. Inconsistency of the maximum parsimony method when the rate of nucleotide substitution is constant. J Mol Evol. 1994;39:210–8.

Tamura K, Nei M, Kumar S. Prospects for inferring very large phylogenies by using the neighbor-joining method. Proc Natl Acad Sci USA. 2004;101:11030–5.

Tamura K, Peterson D, Peterson N, Stecher G, Nei M, Kumar S. MEGA5: molecular evolutionary genetics analysis using maximum likelihood, evolutionary distance, and maximum parsimony methods. Mol Biol Evol. 2011;28:2731–9.

Tavaré S. Some probabilistic and statistical problems in the analysis of DNA sequences. Lect Math Life Sci. 1986;17:57–86.

Thompson JD, Higgins DG, Gibson TJ. CLUSTAL W: improving the sensitivity of progressive multiple sequence alignment through sequence weighting, position-specific gap penalties and weight matrix choice. Nucleic Acids Res. 1994;22(22):4673–80.

Vidyavijayan KK, Hassan S, Precilla LK, Ashokkumar M, Chandrasekeran P, Swaminathan S, Hanna LE. Biased nucleotide composition and differential codon usage pattern in HIV-1 and HIV-2. AIDS Res Hum Retrovir. 2017;33(3):298–307. https://doi.org/10.1089/AID.2015.0320. Epub 2016 Oct 19. PMID: 27599904.

Visseaux B, Damond F, Matheron S, Descamps D, Charpentier C. HIV-2 molecular epidemiology. Infect Genet Evol. 2016;46:233–40.

Whelan S, Goldman N. A general empirical model of protein evolution derived from multiple protein families using a maximum-likelihood approach. Mol Biol Evol. 2001;18:691–9.

Whelan S, Liò P, Goldman N. Molecular phylogenetics: state-of-the-art methods for looking into the past. Trends Genet. 2001;17:262–72.

Woolhouse ME, Gowtage-Sequeria S. Host range and emerging and reemerging pathogens. Emerg Infect Dis. 2005;11:1842–7.

Woolhouse ME, Webster JP, Domingo E, Charlesworth B, Levin BR. Biological and biomedical implications of the co-evolution of pathogens and their hosts. Nat Genet. 2002;32:569–77.

Worobey M, Santiago ML, Keele BF, Ndjango JBN, Joy JB, Labama BL, Dhed'a BD, Rambaut A, Sharp PM, Shaw GM, Hahn BH. Origin of AIDS: contaminated polio vaccine theory refuted. Nature. 2004;428:820.

Yan N, O'Day E, Wheeler LA, Engelman A, Lieberman J. HIV DNA is heavily uracilated, which protects it from autointegration. Proc Natl Acad Sci. 2011;108:9244–9.

Yang Z. Maximum-likelihood estimation of phylogeny from DNA sequences when substitution rates differ over sites. Mol Biol Evol. 1993;10:1396–401.

Yang Z. Maximum likelihood phylogenetic estimation from DNA sequences with variable rates over sites: approximate methods. J Mol Evol. 1994;39(3):306–14.

Yang Z. Computational Molecular Evolution. Oxford, UK: Oxford University Press; 2006.

Yang Z, Rannala B. Bayesian phylogenetic inference using DNA sequences: a Markov Chain Monte Carlo method. Mol Biol Evol. 1997;14:717–24.

Yang Z, Rannala B. Molecular phylogenetics: principles and practice. Nat Rev Genet. 2012;13:303–14.

Zhang J, Nei M. Accuracies of ancestral amino acid sequences inferred by the parsimony, likelihood, and distance methods. J Mol Evol. 1997;44(1):S139–46.

4 Codon Analysis

In this chapter, we delve into the biological underpinnings and statistical metrics of codon usage bias (CUB). Various methodologies exist to quantify CUB, each rooted in distinct theoretical frameworks, biological mechanisms, and codon characteristics; however, this discussion focuses on a select few. We also examine case studies of various viruses to demonstrate how insights into CUB can deepen our understanding of viral evolution, pathogenicity, virus-host interactions, and other key facets of viral life cycles.

Protein-coding genes consist of DNA or RNA sequences that encode functional proteins, with codons made up of nucleotide triplets (Anisimova and Kosiol 2009). The genetic code maps these codons to amino acids, meaning that during protein synthesis, codons are translated into the corresponding amino acids (Anisimova and Kosiol 2009; Cannarozzi and Schneider 2012). While synonymous codons encode the same amino acid and are thus functionally equivalent, they are not used uniformly across genes. This preference for certain codons over others is known as CUB (Cannarozzi and Schneider 2012).

CUB can vary between genes or within a single gene, reflecting the organism's evolutionary and genomic characteristics. For example, CUB can vary among individuals of the same species exposed to different environmental conditions due to differing selection pressures. Additionally, CUB influences gene expression regulation; genes with frequently used codons tend to be expressed at higher levels. Research on CUB spans gene expression regulation, evolutionary biology, and biodiversity and is particularly useful regarding viruses and bacteria, which have rapid generation times. Recently developed statistical methods have enhanced our ability to analyze CUB, improving our understanding of gene regulation and evolutionary dynamics.

CUB is especially significant in viruses due to its impact on host interactions. Viruses, characterized by rapid evolution and high mutation rates, employ various strategies to optimize interactions with their hosts. Adaptive evolution in viruses often involves modulating gene expression to enhance replication efficiency within host cells. CUB is integral to this process, as viruses often use the host's preferred codons to improve translation efficiency. Additionally, codons in cis-acting RNA elements, which regulate gene expression, play a role in viral adaptation. Understanding viral CUB can offer crucial insights into viral biology and host interactions, potentially guiding the development of preventive and therapeutic strategies against viral infections. For instance, insights into viral CUB can inform genome editing strategies and antiviral drug development. Decreasing the frequency of specific codons in viral genomes could inhibit viral survival, while targeting specific codons with reverse transcriptase inhibitors could con-

H. S. Son, M. Cho, *Viroinformatics and Computational Virology in Public Health*,
https://doi.org/10.1007/978-981-96-5757-5_4

trol viral translation. Furthermore, interactions between viruses and RNA-binding proteins (RBPs) influenced by codon usage can be targeted by RNA synthesis inhibitors to inhibit viral replication. Thus, CUB provides valuable information for both understanding viral biology and developing practical applications, such as genome manipulation and antiviral therapies.

4.1 Codon Usage Indices

The genome sequence serves as a blueprint for gene products, offering insights into an organism's functions and evolution. Therefore, investigating codon usage patterns and identifying the selective forces shaping their evolution are vital for genomic research. Analyzing CUB has significant practical implications, particularly for understanding virus evolution and host interactions. The codon bias of the host organism can greatly affect the efficiency of viral protein production, which is essential for viral survival (Kane 1995; Gustafsson et al. 2004). As a result, viruses that rely on host machinery for translation must optimize their codon usage to align with the host's preferences (Wong et al. 2010; Liu et al. 2011). Research on attenuated viral vaccines has shown that substituting optimal codon pairs with synonymous codons in the viral genome can be effective in reducing the rate of viral protein synthesis and enhancing vaccine efficacy (Mueller et al. 2006; Coleman et al. 2008). Thus, a precise and insightful analysis of codon usage patterns is crucial for both research and practical applications. To grasp the causes and consequences of CUB, a quantitative analysis of synonymous codon usage within genes or genomes is essential. Various methods have been developed to measure CUB, and the statistical measures derived from these methods can elucidate the biological factors influencing codon bias in the organism under study (Roth et al. 2012; Cannarozzi and Schneider 2012). The following sections introduce codon usage indices, which are briefly summarized in Table 4.1.

4.1.1 Base Composition

Base composition, typically expressed as GC content, is a fundamental feature of the genome that influences codon usage, DNA methylation, speciation, genome organization, and phylogenetic analysis (Sueoka 1962; Costantini et al. 2006, 2007; Vavouri and Lehner 2011; Eyre-Walker and Hurst 2001). GC composition, which refers to the proportion of guanine (G) and cytosine (C) bases in a DNA or RNA molecule, can be assessed across the entire genome or within specific regions such as a domain, gene, gene cluster, or non-coding segment (Knight et al. 2001). Typically expressed as a percentage, GC content can also be reported as a ratio (G + C ratio or GC ratio). GC content is often preferred over other measures of base composition due to its significant biological implications. Its strong association with genome stability, gene expression levels, codon bias, and evolutionary pressures makes it a valuable indicator in genomic studies. Moreover, GC content directly influences the melting temperature of DNA, which has practical applications in molecular biology techniques, such as PCR and sequencing. Additionally, the variation in GC content among different organisms serves as an important factor in phylogenetic classification and comparative genomics (Sueoka 1962; Eyre-Walker and Hurst 2001). It can be described at three levels: overall GC content, local GC composition, and the G/C or A/T ratio within a single DNA strand (Wan et al. 2004). In living organisms, genomic GC content varies between 25% and 75% overall and between 7% and 95% within individual genes (Lobry and Sueoka 2002; Wan et al. 2004). Local GC composition is categorized according to the positions of codons, with GC1, GC2, and GC3 indicating the GC content at the first, second, and third positions, respectively. According to Watson-Crick base pairing rules, the overall GC content remains consistent between the plus and minus strands of the DNA sequence (Watson and Crick 1953; Wan et al. 2004). By providing insights into genetic and evolutionary processes,

Table 4.1 Key features of codon usage indices

Index	Interpretation	Significance	Applications
Relative synonymous codon usage (RSCU)	Quantifies how frequently a codon is used relative to the expected frequency if all synonymous codons were used equally	Indicates codon usage bias; values >1 suggest a codon is used more often than expected, while values <1 indicate less frequent use.	Studies of gene expression optimization, evolutionary biology, and host-virus adaptation
Frequency of optimal codons (FOP)	Measures the proportion of optimal codons (preferred by highly expressed genes) in a gene	Higher FOP values indicate that a gene is more likely to be highly expressed and efficiently translated.	Predicting gene expression levels, optimizing genes for expression in heterologous systems, and studying translational selection
Codon bias index (CBI)	Quantifies the degree to which a gene uses a subset of preferred codons compared to a reference set of highly expressed genes	Higher CBI values reflect stronger codon bias and potential for higher expression.	Gene expression studies, synthetic biology for gene design, and evolutionary studies of selection pressures on codon usage
Codon adaptation index (CAI)	Measures how well the codon usage of a gene matches the preferred codon usage of highly expressed genes in a reference set	Higher CAI values suggest better adaptation of the gene for translation efficiency in the reference organism.	Optimizing genes for heterologous expression, studying gene expression and evolution, and synthetic biology
Effective number of codons (ENC)	Quantifies the degree of codon usage bias, with values ranging from 20 (extreme bias) to 61 (no bias)	Lower ENC values indicate stronger codon bias and potential selective pressures.	Evolutionary biology, studying translational selection, and analyzing codon usage patterns across different organisms
Intrinsic codon deviation index (ICDI)	Measures the deviation of a gene's codon usage from a random expectation, considering both synonymous and non-synonymous codons	ICDI provides insights into the intrinsic factors affecting codon usage beyond translational selection.	Studies of molecular evolution, gene expression regulation, and mutational biases
Evolutionary rate (ER)	Quantifies the rate at which a gene or genome accumulates mutations over time	Higher evolutionary rates indicate faster changes and potential adaptation to environmental pressures.	Phylogenetics, studying the molecular evolution of genes and genomes, and understanding the dynamics of viral evolution and adaptation

GC content continues to serve as a cornerstone in understanding genome biology and its functional implications across diverse species. The percentage of GC content is calculated as follows:

$$GC\% = \frac{\text{Number of G and C bases}}{\text{Total number of bases}} \times 100$$

where GC% is the GC content percentage. GC ratios within genomes can vary significantly. Genes generally exhibit a higher GC content than that of the overall genome and GC-poor regions where genes are less common (Bernardi 1995; Zoubak et al. 1996). Additionally, GC content differs between organisms (Sueoka 1961). Although the evolutionary drivers of these variations are not fully understood, factors such as selection, mutational bias, and DNA repair mechanisms associated with recombination are believed to influence GC content in different species (Ellegren et al. 2003). Given its distinct interspecies differences, GC content is often used for species classification, particularly when traditional systems are inadequate. For instance, the GC ratio has been employed in prokaryote taxonomy to suggest various taxonomic groupings, as recommended by the Ad Hoc Committee on Reconciliation of Approaches to Bacterial Systematics. Due to the constraints of the genetic code, achieving a genome with a GC content nearing 0% or 100% is virtually impossible. For example, *Plasmodium falciparum*, which has an exceptionally low GC content (~20%), is typically described as AT-rich rather than GC-poor (Bisen et al. 2012). CUB often correlates with GC composition, with extreme GC content values usually associated with highly biased codon usage (Wan et al. 2004). Thus, analyzing the distributions and patterns of GC content and CUB can offer valuable insights into the evolutionary characteristics of species.

4.1.2 Relative Synonymous Codon Usage

The relative synonymous codon usage (RSCU) index, introduced by Sharp and Li (1986, 1987), quantifies codon bias by assigning a numerical value to each codon, reflecting its usage relative to a random distribution of synonymous codons (Sharp and Li 1986, 1987). The RSCU value for a codon is determined by dividing its observed frequency by the frequency expected if all synonymous codons for the same amino acid were used equally. For instance, if an amino acid is encoded by four synonymous codons, each codon would be expected to appear 25% of the time in the absence of bias. The RSCU for each codon in a nucleotide sequence is calculated using the following formula:

$$\mathrm{RSCU}_{ij} = \frac{\text{Observed frequency of the codon}}{\text{Expected frequency of the codon}} = \frac{X_{ij}}{\frac{1}{n_\mathrm{i}}\sum_{j=1}^{n_i} X_{ij}}$$

where $RSCU_{ij}$ is the RSCU value for codon j encoding amino acid I, X_{ij} represents the observed number of occurrences of codon j for amino acid I, n_i denotes the number of synonymous codons for amino acid I (ranging from one to six), and $\sum_{j=1}^{n_i} X_{ij}$ is the total number of occurrences of all synonymous codons for amino acid i. The RSCU values of individual codons are used to measure the adaptation of a gene to biased codon usage (Sharp and Li 1987). RSCU values indicate the ratio between the observed frequency of a particular codon and the frequency expected if there were no CUB. When there is no bias, the RSCU value is 1. Codons used more frequently than expected will have an RSCU value greater than 1, while those used less frequently than expected will have an RSCU value less than 1 (Roth et al. 2012).

RSCU values are particularly valuable for normalizing codon usage patterns, facilitating comparisons of codon bias across different genes or organisms. By analyzing RSCU values, researchers can identify preferred codons in specific organisms or tissue types, gaining insights into various biological processes. Several factors can influence CUB, and RSCU values assist in studying these factors:

1. *Gene expression levels:* Highly expressed genes often exhibit stronger CUB, favoring codons that match abundant tRNAs, thereby enhancing translation efficiency.
2. *GC content:* Organisms with varying genomic GC content may exhibit biases towards codons with higher GC content.
3. *Evolutionary adaptation:* Codon usage can reflect evolutionary pressures, such as selection for translational accuracy or adaptation to specific environmental conditions.
4. *Horizontal gene transfer:* RSCU values can help identify genes acquired through horizontal gene transfer, which may show different codon usage patterns compared to native genes.

RSCU values can also be visualized using heatmaps or other graphical formats to illustrate CUB across multiple genes or genomes. Such visualizations help detect trends and anomalies in codon usage that might be related to functional or evolutionary processes. In summary, the RSCU index provides a normalized measure of CUB, revealing how frequently specific codons are used compared to a random distribution. It is a valuable tool for investigating the factors influencing codon bias and understanding its evolutionary and functional implications.

4.1.3 Frequency of Optimal Codons

The frequency of optimal codons (FOP) is a metric used in molecular biology and genetics to quantify the preference of a species or organism for specific synonymous codons during translation (Ikemura 1981). This preference relates to the efficiency and accuracy of protein synthesis, as optimal codons are translated more efficiently and accurately by the cellular machinery, often correlating with the abundance of corresponding tRNA molecules. FOP is defined as the ratio of the number of times optimal codons are used to the total number of synonymous codons for a particular amino acid in a gene. A higher FOP value signifies greater usage of preferred codons, which is often linked to highly expressed genes. This metric provides insights into codon usage patterns in relation to gene expression and protein synthesis.

Conceptually, FOP is based on the idea that natural selection favors codons that enhance the efficiency and accuracy of protein synthesis, particularly in highly expressed genes where translational efficiency is crucial. This selective pressure results in a non-random usage of synonymous codons, with certain codons being preferred over others even though they encode the same amino acid. Understanding FOP provides insights into the evolutionary pressures on genes and the strategies organisms use to optimize gene expression. For example, genes that are highly expressed in rapidly dividing cells or in response to specific environmental conditions may show a higher FOP, indicating an adaptation to maximize protein production speed as well as accuracy.

To calculate the FOP for a gene, first, the optimal codons for the organism are determined; these are often available in organism-specific codon usage tables. Then, the total number of codons and the number of optimal codons in the gene are counted. FOP is computed by dividing the number of optimal codons by the total number of codons in the gene. The formula for calculating FOP is as follows:

$$\text{FOP} = \frac{\text{Number of optimal codons}}{\text{Total number of codons}}$$

For instance, if a gene contains 300 codons and 180 of these are optimal, the FOP would be 0.6, meaning 60% of the codons in this gene are optimal. FOP is a useful metric for studying CUB and its effects on gene expression and evolutionary biology. It helps compare gene expression levels, as genes with higher FOP values are generally more highly expressed. Additionally, FOP offers insights into evolutionary pressures, as natural selection may favor optimal codons in highly expressed genes. In biotechnology, FOP assists in designing synthetic genes for optimal expression in host organisms by selecting preferred codons. In summary, FOP measures the preferential use of optimal codons, reflecting the relationship between genetic coding and translational efficiency, and is valuable for examining CUB and its implications.

4.1.4 Codon Bias Index

The codon bias index (CBI), introduced by Bennetzen and Hall in 1982, quantifies the extent to which a gene favors preferred codons over alternative synonymous ones (Bennetzen and Hall 1982). This index helps measure CUB in a gene relative to the most biased gene in the genome. Preferred codons are translated more efficiently and accurately due to their corresponding tRNA molecules being more abundant or effective at pairing with mRNA during translation. Therefore, a gene with a high CBI score heavily utilizes preferred codons. The CBI assesses the extent to which a gene uses the most preferred codons compared to the expectation with random codon usage. The calculation involves comparing the observed frequency of preferred codons in a gene to the expected frequency if codon usage were random. The index ranges from 0 to 1, where 0 signifies no bias (random use of all synonymous codons) and 1 indicates maximal bias (exclusive use of preferred codons). Higher CBI values denote a stronger preference for optimal codons. The formula for calculating CBI is as follows:

$$\text{CBI} = \frac{Nop - Nref}{N - Nref}$$

where *Nop* is the number of occurrences of optimal codons in the gene, *Nref* is the number of occurrences of optimal codons in a reference set of highly expressed genes, and *N* is the total number of codons in the gene.

CBI is useful for studying gene expression levels and evolutionary pressures. Highly expressed genes often have higher CBI values, as preferred codons enhance translational efficiency and accuracy. In contrast, genes with lower expression or those less crucial for rapid responses may have lower CBI values, reflecting more random codon usage. In evolutionary biology, CBI can reveal how selection pressures influence codon usage patterns. For instance, organisms requiring rapid protein production may evolve higher CBI values for essential genes. Comparing CBI values across species provides insights into their evolutionary strategies and adaptations related to gene expression. Overall, CBI is a comprehensive metric to evaluate codon preference within genes, elucidating the relationship between codon usage, gene expression, and evolutionary pressures. It is valuable for assessing CUB, guiding both biological research and biotechnological applications.

4.1.5 Codon Adaptation Index

The codon adaptation index (CAI), introduced by Sharp and Li in 1987, is a widely used metric in molecular biology and genetics that measures how well a gene's codon usage aligns with the preferred codon usage of highly expressed genes in an organism (Sharp and Li 1987). This index assesses the degree to which a gene's codon usage reflects patterns known to enhance translation efficiency and accuracy. CAI quantifies the similarity between a gene's codon usage and the codon usage patterns of highly expressed genes in a reference set. The reference set represents the optimal codon usage pattern for the organism and is typically derived from genes with high expression levels. Each codon in the gene of interest is assigned a relative adaptiveness value, calculated as the ratio of its observed frequency in the reference set to the maximum frequency of synonymous codons for the same amino acid. These values range from 0 to 1, with 1 representing the most frequently used, or "optimal," codon. The formula for calculating the CAI is as follows:

$$\omega_i = \frac{f_i}{f_{\max}}$$

where ω_i is the relative adaptiveness value for codon i, f_i is the frequency of codon i in the reference set, and f_{max} is the frequency of the most frequent synonymous codon for that amino acid. The CAI for a gene with n codons is calculated using the geometric mean of these weights as follows:

$$\text{CAI} = \left(\prod_{n}^{i=1} \omega_i \right)^{\frac{1}{n}}$$

To compute the CAI, for a gene, the relative adaptiveness values of all codons in the gene are multiplied, and then the n_{th} root of this product is taken, where n is the number of codons in the gene. The resulting CAI, value ranges from 0 to 1, with higher values indicating a closer alignment to the codon usage patterns of highly expressed genes in the reference set. A CAI of 1 signifies perfect codon optimization, meaning the gene uses only the most preferred codons, while a value near 0 reflects a more random or less optimal codon usage pattern.

The CAI is a valuable tool for several reasons. It enables researchers to predict a gene's potential expression level based on its codon usage. Genes with higher CAI values are likely to be translated more efficiently, as their codon usage aligns closely with that of highly expressed genes. This makes CAI particularly useful in biotechnology and synthetic biology, which often involves optimizing genes for expression in target organisms. For instance, when designing synthetic genes for bacteria like *E. coli*, researchers can adjust codon usage to achieve a higher CAI and thus enhance protein production. Additionally, CAI offers insights into evolutionary pressures on codon usage. By comparing CAI values across different species or conditions, researchers can infer how natural selection and mutational biases have influenced codon usage patterns, revealing details about gene function, expression levels, and evolutionary history. In summary, CAI is an effective metric for evaluating how well a gene's codon usage matches the optimal codon usage of highly expressed genes, providing valuable insights into gene expression potential and evolutionary pressures.

4.1.6 Effective Number of Codons

The effective number of codons (ENC), introduced by Wright in 1990, quantifies the degree of CUB in a gene (Wright 1990). Unlike other metrics, ENC estimates how much a gene's codon usage deviates from a random, uniform distribution of synonymous codons. It measures how many codons are used by a gene relative to the total number of possible codons. ENC values range from 20 to 61: an ENC value of 20 indicates extreme codon bias, where only one codon is used for each amino acid, while an ENC value of 61 suggests no bias, with all synonymous codons used equally, reflecting random codon usage. The formula for calculating ENC is as follows:

$$\mathrm{ENC} = 2 + \frac{9}{F_2} + \frac{1}{F_3} + \frac{5}{F_4} + \frac{3}{F_6}$$

where F_k represents the average homozygosity (the probability that two randomly chosen codons for an amino acid are the same) for amino acids with k synonymous codons.

ENC is calculated based on the frequencies of synonymous codons within a gene, considering the observed usage of each codon relative to its synonymous alternatives. The process involves determining the codon frequencies for each amino acid, computing their homozygosity values, and then applying the ENC formula to obtain the gene's overall ENC. A key advantage of ENC is its adjustment for amino acid composition and gene length, which makes it a robust measure for comparing CUB across genes with varying lengths and amino acid compositions.

ENC values are less affected by gene length and overall base composition compared to metrics such as GC content or the frequency of individual codons, providing a standardized metric for CUB. Practically, ENC is used to infer selective pressures acting on a gene. Genes with low ENC values typically exhibit strong CUB, often linked to high expression levels and pronounced translational selection. Such genes are likely to use codons corresponding to the most abundant tRNAs in the cell, thereby optimizing translation efficiency and accuracy. In contrast, genes with high ENC values show minimal CUB, suggesting that translational selection has less impact. These genes might be expressed at lower levels or face weaker selective constraints regarding codon usage. Additionally, ENC can reveal other influences on codon usage, such as patterns of mutational bias, where certain nucleotides are more prone to mutation than others. By comparing ENC across different organisms or conditions, researchers can assess the relative contributions of natural selection and mutational biases to codon usage patterns. In summary, ENC is a valuable metric for evaluating CUB, offering insights into gene expression, evolutionary pressures, and comparative genomics. It quantifies how much a gene's codon usage deviates from uniform usage, with lower values indicating greater bias.

4.1.7 Intrinsic Codon Deviation Index

The intrinsic codon deviation index (ICDI), introduced by Freire-Picos et al. in 1994, assesses CUB within genes (Freire-Picos et al. 1994). Unlike other metrics, ICDI evaluates how a gene's codon usage deviates from a uniform distribution of synonymous codons without relying on predefined optimal codons or genome-wide codon frequencies. To calculate ICDI, the observed frequencies of each codon are compared to their expected frequencies, assuming uniform usage. The formula for ICDI quantifies this deviation as follows:

$$\mathrm{ICDI} = \sum \frac{\left(f_i - e_i\right)^2}{e_i}$$

where f_i represents the observed frequency of the ith codon and e_i represents the expected frequency under uniform usage. The resulting ICDI value indicates the extent to which the actual codon usage deviates from a uniform distribution.

A low ICDI value indicates that a gene's codon usage closely aligns with a uniform distribution, suggesting minimal intrinsic bias. Conversely, a high ICDI value reflects a substantial deviation from uniformity, signaling significant CUB, likely driven by evolutionary pressures, such as natural selection for translational efficiency and accuracy, mutational biases, and other selective constraints.

One of ICDI's key advantages is its applicability across various organisms and genes without the need for species-specific information about optimal codons. This characteristic enables researchers to perform comparative studies of CUB in a wide range of organisms, including those with poorly characterized genomes or limited genomic resources. By analyzing ICDI values, researchers can gain insights into how evolutionary and ecological factors shape codon usage patterns. This includes understanding the relative impacts of selective pressures and mutational biases, which can illuminate the molecular evolution of genes and the adaptive strategies of organisms within their environments.

In summary, ICDI provides a comprehensive assessment of CUB based on the intrinsic properties of the genetic code and amino acid sequences. It offers a robust framework for exploring CUB that transcends species boundaries and genomic complexities, enhancing our understanding of the evolutionary forces and functional constraints that influence codon usage.

4.1.8 Evolutionary Rate

The evolutionary rate for codon usage (ER) is a key metric in evolutionary biology that assesses how codon usage patterns evolve over time across different species. The concept of ER was introduced by Grantham et al. in 1981, motivated by the observation that codon usage tends to be more similar among closely related species and shows more pronounced variation over longer evolutionary distances (Grantham et al. 1981). Thus, ER is a useful indicator of evolutionary distance. ER quantifies the rate at which genetic changes accumulate in a given sequence, reflecting both evolutionary distances and selective pressures on genes. This correlation with evolutionary distance helps in understanding historical relationships and divergence times among species.

To calculate ER, researchers typically compare the codon usage frequencies of homologous genes or gene sets across species. The calculation involves determining the number of substitutions per site (K) between sequences and estimating the divergence time (T) since their most recent common ancestor. The formula for ER is expressed as follows:

$$ER = \frac{K}{2T}$$

where K represents the number of substitutions per site and T is the divergence time in years or other appropriate units. For example, if two sequences have an average of 0.01 substitutions per site and diverged one million years ago,

$$ER = \frac{0.01}{2 \times 1,000,000} = 5 \times 10^{-9.}$$

This indicates that there have been 5×10^{-9} substitutions per site per year.

Various factors impact ER, including mutation rates, selection pressures, genetic drift, and environmental changes. Synonymous substitutions, which alter codon usage without changing the amino acid, are generally considered neutral regarding selection, whereas non-synonymous substitutions can influence protein function and are subject to selective pressures. By focusing on codon usage patterns rather than solely on nucleotide changes, ER provides a refined understanding of how translational selection and mutational biases influence the optimization of coding sequences during evolution. ER is extensively used in evolutionary biology to gain insights into genetic evolution and divergence among species. In phylogenetics and the molecular clock hypothesis, ER is crucial for constructing phylogenetic trees and estimating divergence times, assuming a relatively constant mutation rate. Thus, it provides perspectives on evolutionary history and temporal relationships among organisms. Additionally, ER is valuable in studying adaptive evolution, identifying genes with rapid changes in codon usage patterns and, thus, highlighting adaptations to new environments or functional shifts. In comparative genomics, ER facilitates the comparison of evolutionary rates across species and within genes, helping to identify conserved regions indicating functional constraints and divergent regions suggesting adaptive changes or innovations. Overall, ER is a versatile metric that links phylogenetic studies, adaptive evolution, and genomic dynamics, offering quantitative measures of genetic change over time. High ER values indicate rapid genetic changes, potentially due to high mutation rates or strong positive selection, while low ER values suggest slower changes, likely due to purifying selection or low mutation rates. In summary, ER serves as an essential metric for understanding the evolution of codon bias, the adaptation of gene expression mechanisms, and the historical relationships among organisms. By capturing how codon preferences shift over evolutionary time, ER complements other codon usage indices and provides deeper insights into the molecular evolution of coding sequences (Cannarozzi and Schneider 2012).

4.2 Evolution of Codon Usage in Viruses

A virus relies on the cellular machinery of its host organism to complete its life cycle, including the processes of entry, replication, and assembly. Since most viral genomes do not encode tRNAs, viral protein translation depends on the host's tRNA pool. Consequently, viruses tend to evolve their codon usage to align with that of their hosts to efficiently use the host's resources (Zhou et al. 1999; Rocha 2004; Chaney and Clark 2015). The genetic code is characterized by the unequal usage of synonymous codons (coding for the same amino acid) due to its degeneracy (Crick, 1961). This redundancy allows for efficient and accurate protein production while preserving the amino acid sequence (Robinson et al. 1984; Spencer and Barral 2012). A closer match in codon usage patterns between a virus and its host species leads to easier viral replication. Studies have suggested that the extent and frequency of CUB in viruses and their hosts can influence viral survival, adaptation, and evasion of the host's immune system (Carbone 2008). CUB is common in viruses, and the codon usage of viruses and their hosts often co-evolves (Berkhout et al. 2002; Jenkins and Holmes 2003; Bahir et al. 2009; Lobo et al. 2009; Liu et al. 2011). This evidence supports the idea that efficient replication requires viruses and their hosts to have similar codon usage patterns, enabling the sharing of tRNA pools (Chaney and Clark 2015; Komar 2016). Viruses can have varying host ranges, which may differ depending on their species or subspecies. Some viruses, such as arboviruses, have a broad host range, infecting mammals, birds, and insects (Weaver and Barrett 2004), while others, such as Hepatitis B virus (HBV) or Human Immunodeficiency Virus (HIV), are highly host-specific, primarily infecting humans and certain primates (Kramvis and Kew 2007; Sharp and Hahn 2011). The codon usage of viruses with broad host ranges may be adapted to the tRNA pools of each host, meaning that viral progeny resulting from replication and assembly will exhibit different codon usage patterns depending on their host species (Bahir et al. 2009). In other words, viruses from the same genus or species may show variations in codon usage based on their host, which could be influenced by the host's tRNA composition.

Given that the codon usage of viral genes may be subject to host selection pressures, particularly after interspecies transmission, analyzing it in relation to host specificity could provide insights into viral evolution, adaptive host responses, and viral infection control methods (Wong et al. 2010; Mani et al. 2011). For influenza A viruses, subtypes are defined by the types of hemagglutinin (HA) and neuraminidase (NA), which determine their broad host range. HA and NA are classified into subtypes H1 to H18 and N1 to N11, respectively, with each subtype showing different levels of infectivity for humans, birds, pigs, bats, and other animals (Centers for Disease Control and Prevention, National Center for Immunization and Respiratory Diseases (NCIRD), https://www.cdc.gov/flu/about/viruses/types.htm, March 30, 2023). While the precise manner in which host range is determined in influenza A viruses remains uncertain, it is clear that the specificity of HA's receptor binding plays a critical role, particularly in cross-species infections, which is another important epidemiological feature of the virus. One significant aspect of determining the host range of influenza A viruses is the role of pigs as a "mixing vessel" (Scholtissek 1990; Ito et al. 1998). Pigs' respiratory cells contain both α2,3-linked (avian-type) and α2,6-linked (human-type) sialic acid receptors (Graaf and Fouchier 2014), allowing the virus to bind to both types of receptors. Given the impact of HA protein mutations on cross-species infection, the variety of receptor types in pigs may serve as a major source of new viral quasi-species with diverse mutations that can infect multiple hosts with various HA types.

Previous studies on human RNA viruses have shown that mutational pressure, rather than translational selection, is a key factor driving CUB (Sharp et al. 1993; Jenkins and Holmes 2003). In their 2010 study, Wong et al. investigated the codon usage of influenza viruses in mammalian species using RSCU values and correspondence analysis (CA). By analyzing codon usage patterns, they revealed evolutionary trends, categorizing viruses based on their hosts and subtypes. Their study also validated translation- and mutation-related models of CUB by revealing how codon usage changed over time in human influenza viruses. They suggested that viral evolution is mainly driven by base composition, with the host's selective pressure on GC content playing a significant role (Wong et al. 2010).

While previous studies have identified common codon biases in influenza viruses, the effect of host selective pressures on codon usage across avian, swine, and human influenza viruses, especially in relation to host switching, remains unclear. Understanding pigs' role as intermediate hosts in cross-species infection by influenza A viruses could inform preparations for the emergence of new influenza strains in human populations, critical for designing effective interventions to mitigate future viral outbreaks.

The following case study illustrates how CUB analysis can be applied to elucidate various characteristics of viruses, including their evolution, host range, cross-species infection, and virus-host interactions (Cho 2019). This section focuses on influenza A virus analysis. For other case studies involving codon analysis of different viruses, please refer to "Part III. Applications and Case Studies".

4.3 Case Study: Codon Usage Analysis of Influenza A Virus

This study involved a codon-based evolutionary analysis of influenza A viruses, focusing on gene sequences encoding host-specific viral proteins. GC content and RSCU were computed to discern codon usage patterns across virus species based on their respective host organisms. While GC content and codon usage are critical sequence features influencing gene expression, the molecular mechanisms governing their effects remain only partially understood (Shackelton et al. 2006; Barahimipour et al. 2015). Viruses often align their codon usage preferences with their hosts' biases, reflecting their reliance on the host's cellular machinery for replication, transcription, and translation. Consequently, the codon usage indices of host-specific viral proteins provide valuable insights into the genetic and evolutionary interplay between viruses and their hosts, exemplifying co-evolution and adaptive processes. Moreover, these host-specific codon usage patterns have predictive potential for infection probabilities in specific hosts, whether primary or reservoir species. Thus, codon usage patterns serve as critical markers of the interspecies transmissibility and reproductive success of viral progeny.

In this study, the HA gene encoding the receptor-binding protein of influenza A viruses was analyzed to infer codon usage patterns relative to the species of origin. The relative contributions of GC content and codon usage to the translational efficiency of HA genes from the

H1N1, H3N2, and H5N1 viral subtypes were evaluated. By comparing GC content and codon usage in gene variants encoding orthologous amino acid sequences across host species, this study assessed codon usage as a determinant of virus-host interactions, translational productivity, and viral adaptability.

Nucleotide composition and codon usage analyses were conducted for the H1N1, H3N2, and H5N1 subtypes isolated from three host species. Changes in codon usage patterns within HA-coding genes, a determinant of host specificity, were evaluated based on host species. Nucleotide sequences for the HA-coding region were retrieved from the NCBI Influenza Virus Database (https://www.ncbi.nlm.nih.gov/genomes/FLU/Database/nph-select.cgi?go=database). The datasets included 450 avian H1N1, 3363 swine H1N1, 16,679 human H1N1, 365 avian H3N2, 2708 swine H3N2, 19,158 human H3N2, 3057 avian H5N1, 27 swine H5N1, and 266 human H5N1 sequences. Maximum, minimum, arithmetic mean, variance, and standard deviation values for GC content in HA-coding genes were calculated for each subtype. Mean RSCU values were derived for each dataset to examine host-specific codon usage variations.

Table 4.2 presents the GC content of HA-coding genes in influenza A viruses from avian, swine, and human hosts. The HA genes of H1N1 and H3N2 viruses from avian hosts exhibited higher mean GC content than those from swine and human hosts. The maximum GC content, variance, and standard deviation were also lower in HA genes from swine and human hosts. (Table 4.2 and Fig. 4.1). The GC content differences between host species were used to infer genetic divergence and cross-species infection risks. The smallest mean GC content differences were observed in human-swine pairs for H1N1 (0.09) and H3N2 (0.21) and avian-swine pairs for H5N1 (0.80). In contrast, the largest differences occurred in avian-swine pairs for H1N1 (2.56),

Table 4.2 GC content comparison of HA-coding genes across avian, human, and swine influenza A viruses

Subtype	Host	*N*	GC contents				
			Mean	Max.	Min.	Var.	STDEV.
H1N1	Avian	450	43.28	44.44	35.92	0.66	0.81
	Swine	3363	40.72	43.80	38.04	0.17	0.42
	Human	16,679	40.81	42.58	37.74	0.09	0.30
H3N2	Avian	365	43.24	45.21	41.67	0.50	0.71
	Swine	19,158	42.04	44.33	40.68	0.16	0.40
	Human	2708	41.83	44.62	27.75	0.25	0.50
H5N1	Avian	3057	41.03	42.54	32.04	0.14	0.37
	Swine	266	41.24	42.06	40.42	0.16	0.40
	Human	27	42.04	44.33	40.68	0.16	0.40
Subtype	Host pair		The difference of GC contents between hosts				
			Mean	Max.	Min.	Var.	STDEV.
H1N1	Avian	Swine	2.56†	0.64‡	2.12†	0.49	0.39
	Avian	Human	2.47	1.86†	1.82	0.57†	0.51†
	Human	Swine	0.09‡	1.22	0.30‡	0.08‡	0.12‡
H3N2	Avian	Swine	1.20	0.88†	0.99‡	0.34†	0.31†
	Avian	Human	1.41†	0.59	13.92†	0.25	0.21
	Human	Swine	0.21‡	0.29‡	12.93	0.09‡	0.10‡
H5N1	Avian	Swine	0.21‡	0.48‡	8.38	0.02†	0.03†
	Avian	Human	1.01†	1.79	8.64†	0.02†	0.03†
	Human	Swine	0.80	2.27†	0.26‡	0.00‡	0.00‡

This table presents a summary of the GC content (mean, maximum, minimum, variance, and standard deviation) in the HA-coding regions of influenza A virus subtypes found in avian, swine, and human hosts. Additionally, pairwise comparisons between host species illustrate the extent of GC content differences for each subtype.N, Number of sequences analysed; † The largest GC content difference observed between host species; ‡ The smallest GC content difference observed between host species

avian-human pairs for H3N2 (1.41), and avian-human pairs for H5N1 (1.01). These findings implicate avian hosts as the primary contributors to increased GC content divergence and swine hosts as potential mediators reducing these differences.

As detailed in Sect. 4.1, codons with RSCU values exceeding 1.0 exhibit positive CUB (abundant codons), while those with values below 1.0 exhibit negative bias (less abundant codons). An RSCU value of 1.0 indicates no CUB (Sharp and Li 1986). Codons with RSCU values above 1.6 are considered overrepresented, while those below 0.6 are underrepresented (Wong et al. 2010). RSCU analysis revealed that swine and human influenza viruses share abundant codons (e.g., ATT[Ile], AAA[Lys], TGT[Cys] for H1N1; CCA[Pro], ACT[Thr], TAC[Tyr], GAC[Asp] for H3N2; and GAT[Asp] for H5N1) and overrepresented codons (e.g., CTA[Leu], GTA[Val], CCA[Pro] for H1N1 and GTT[Val] for H3N2). (Fig. 4.2). These results indicate that GC content and RSCU values are reflective of host specificity and may serve as predictive markers for viral infectivity.

The differences in mean GC content and RSCU values across avian, swine, and human hosts corroborate host-specific viral adaptation. Among the H1N1 and H3N2 subtypes, the smallest GC content differences were observed in swine-human pairs, while avian-swine pairs exhibited the smallest differences for H5N1. Similarly, pairwise RSCU distances were smallest in swine-human pairs for H1N1 and H3N2 and in avian-human pairs for H5N1, affirming the host's influence on viral codon usage (Figs. 4.3, 4.4 and 4.5).

Previous studies have suggested that pigs act as "mixing vessels" for generating human-avian influenza reassortants, as seen during the 1957 and 1968 pandemics (Ito et al. 1998; Zhou et al. 1999). Their findings support the hypothesis that avian influenza viruses adapt to human hosts through mutations or reassortment events within pigs. Conversely, other studies have posited that H5N1 can directly infect humans without involving pigs as intermediaries. In these studies, although direct contact between the infected child and affected chickens could not be definitively established, the findings imply that the virus was transmitted from chickens to the child without the involvement of another mammalian host (Claas et al. 1998). While highly pathogenic H5N1 viruses have occasionally been detected in pigs (Webster et al. 1992; Connor

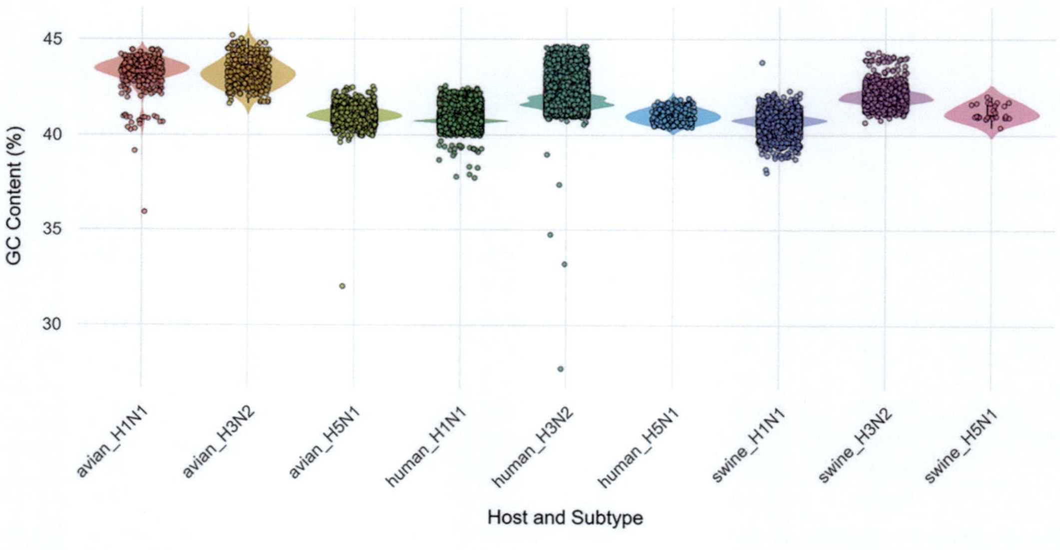

Fig. 4.1 Distribution of GC content in HA-coding genes across influenza A virus subtypes and hosts. Shown are the maximum, minimum, mean, and standard deviation values for each group. The number of sequences analyzed for GC content is as follows: avian H1N1 (450), swine H1N1 (3,363), human H1N1 (16,679), avian H3N2 (365), swine H3N2 (2,708), human H3N2 (19,158), avian H5N1 (3,057), swine H5N1 (27), and human H5N1 (266)

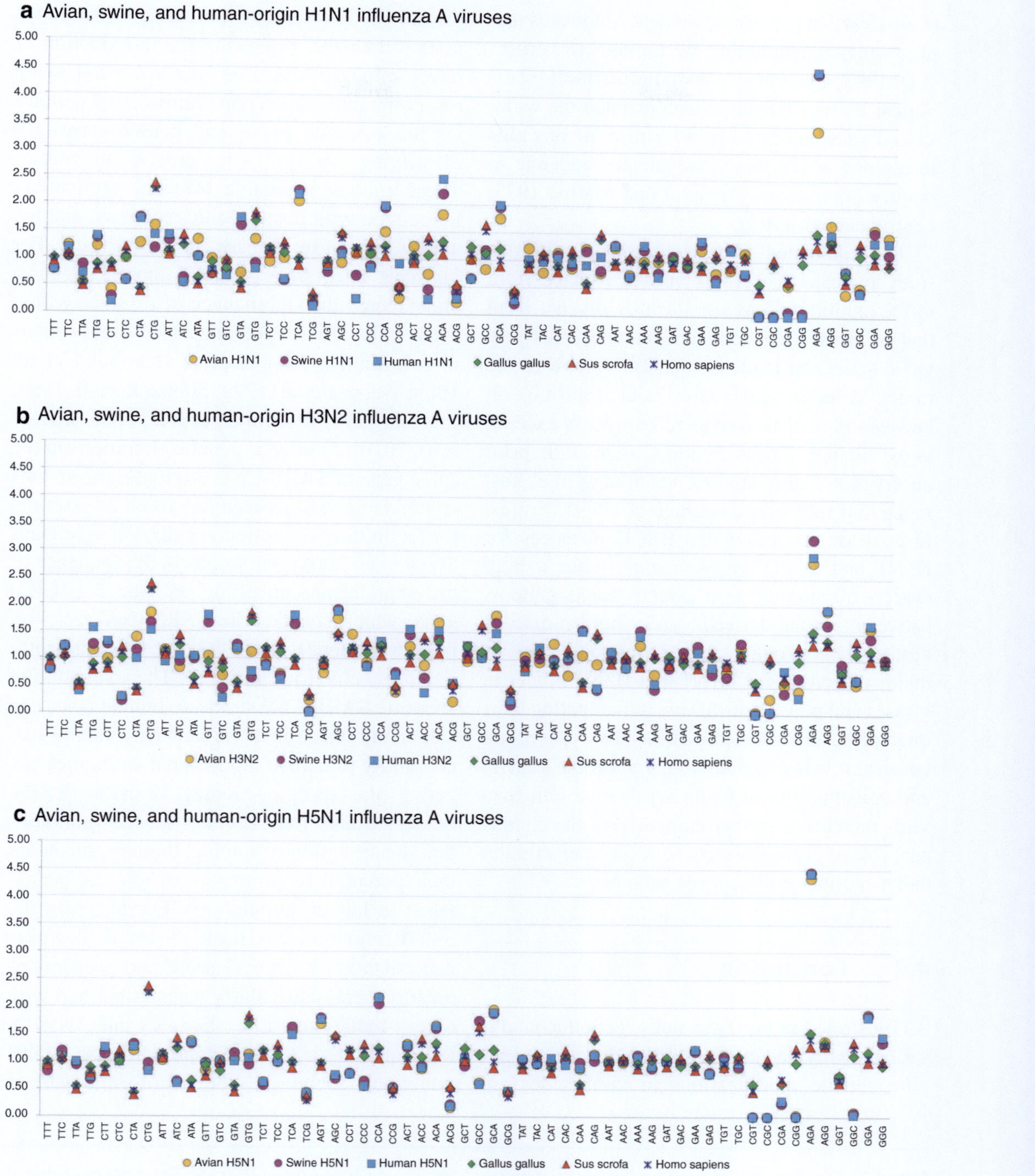

Fig. 4.2 This figure presents the relative synonymous codon usage (RSCU) profiles of the HA-coding regions across various influenza A virus subtypes and their respective host genomes. Each panel illustrates RSCU patterns for viruses derived from avian, swine, and human hosts, in comparison with the codon usage profiles of the corresponding host species. Specifically, panel (a) shows RSCU profiles for H1N1 viruses, panel (b) for H3N2 viruses, and panel (c) for H5N1 viruses. The RSCU values represent the arithmetic mean calculated for each viral group, based on the following number of sequences: avian H1N1 (450), swine H1N1 (3,363), human H1N1 (16,679), avian H3N2 (365), swine H3N2 (2,708), human H3N2 (19,158), avian H5N1 (3,057), swine H5N1 (27), and human H5N1 (266)

et al. 1994), experimental evidence indicates that pig-to-pig transmission is limited (Castrucci et al. 1993; Claas et al. 1994). Furthermore, serological surveys have not demonstrated the widespread transmission of H5N1 viruses among pigs in regions where this virus remains endemic in poultry populations (Masurel and Marine 1973; Castrucci et al. 1993).

The calculated RSCU distances from this study indicate that H5N1 viruses isolated from swine exhibit a CUB significantly distinct from that of viruses isolated from humans. This provides molecular evidence that H5N1 viruses are rarely detected in swine, and transmission between pigs or from pigs to humans is exceedingly limited. These findings align with prior observations and studies examining the host range and infection dynamics of H5N1 viruses. In contrast, the calculated RSCU distances for H1N1 and H3N2 viruses demonstrate a high degree of similarity in codon usage patterns between swine-derived and human-derived viruses. This strongly suggests that pigs serve as major reservoirs for H1N1 and H3N2 influenza viruses and play a critical role in facilitating their interspecies transmission. These observations are consistent with previous research on the genetic and epidemiological features of swine influenza virus infections, further emphasizing the significant role of pigs in the cross-species transmission and evolution of these virus subtypes.

4.4 Conclusion

As illustrated in this case study, genomic analyses have provided compelling evidence that specific genetic modifications in viral genes are pivotal for interspecies transmission from animals to humans. Following the introduction of novel influenza A viruses from avian and swine reservoirs into human populations, viral genomics has emerged as an indispensable tool for identifying viruses with critical mutations. These mutations enable efficient replication, interaction with host-specific receptors, and the ability to cause disease in new hosts. For instance, integrative genomic approaches have underscored the significance of residues within the HA, PA, PB1, PB2, and other viral proteins in determining host tropism (Kawaoka et al. 1989; Webster et al. 1992; Subbarao et al. 1993; Ito et al. 1998; Taubenberger et al. 2005; Bussey et al. 2010). Similarly, genetic alterations in the spike gene of SARS-CoV, originating from bats and civet cats, have been identified as essential for facilitating zoonotic transmission to humans (Tang et al. 2006). Variations in the sequence of the angiotensin-converting enzyme 2 (ACE2) gene, which encodes the SARS-CoV receptor, further influence the efficiency of viral entry into host cells (Li et al. 2005). The adaptation of zoonotic SARS-CoV strains to human hosts can lead to the selection of variants with altered susceptibility to human monoclonal antibodies targeting the receptor-binding domain (RBD). Some variants may develop escape mutations that reduce antibody binding, thereby enhancing their potential for immune evasion and persistence in human populations. For instance, the S494P mutation has been shown to decrease neutralization by convalescent and post-immunization sera, particularly when combined with other mutations like E484K and N501Y (Alenquer et al. 2021). Additionally, the L452R mutation, found in the California variant

Fig. 4.3 Pairwise RSCU distance profiles between viruses and hosts in the H1N1 subtype. This figure illustrates the pairwise comparisons of relative synonymous codon usage (RSCU) distances in the HA-coding regions of H1N1 influenza A viruses and their corresponding host genomes. The panels present RSCU distance patterns for different types of pairings: (**a**) comparisons between different H1N1 virus groups (avian, swine, and human); (**b**) comparisons among host genomes (*Gallus gallus*, *Sus scrofa*, and *Homo sapiens*); (**c**) RSCU distances between avian H1N1 viruses and each host genome; (**d**) RSCU distances between swine H1N1 viruses and each host genome; (**e**) RSCU distances between human H1N1 viruses and each host genome. Each data point represents the codon-specific RSCU distance for a given pairwise comparison, providing insight into codon usage divergence across virus–virus, host–host, and virus–host relationships

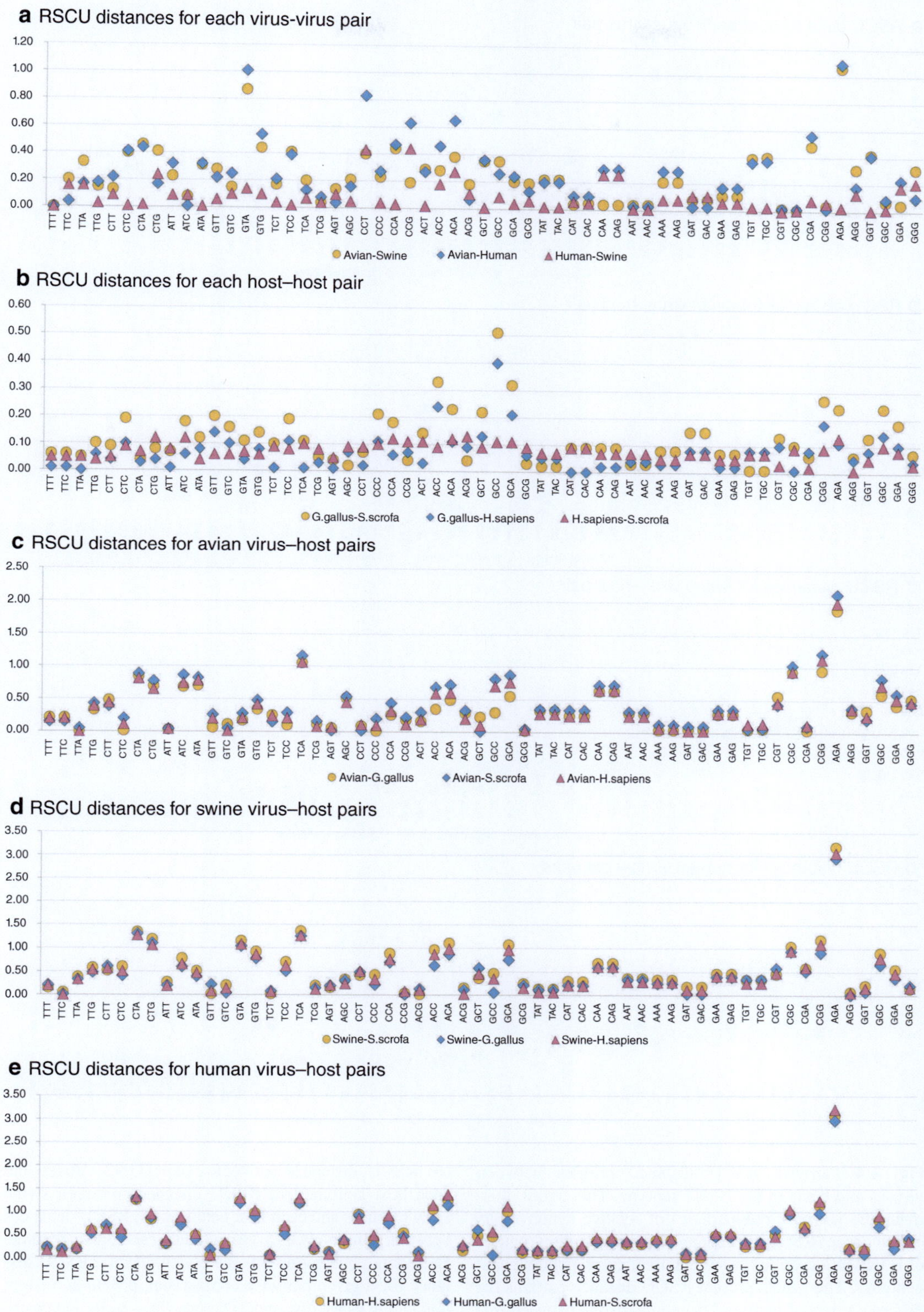
a RSCU distances for each virus-virus pair
1.20
1.00
0.80
0.60
0.40
0.20
0.00
TTT TTC TTA TTG CTT CTC CTA CTG ATT ATC ATA GTT GTC GTA GTG TCT TCC TCA TCG AGT AGC CCT CCC CCA CCG ACT ACC ACA ACG GCT GCC GCA GCG TAT TAC CAT CAC CAA CAG AAT AAC AAA AAG GAT GAC GAA GAG TGT TGC CGT CGC CGA CGG AGA AGG GGT GGC GGA GGG
Avian-Swine
Avian-Human
Human-Swine
b RSCU distances for each host–host pair
0.60
0.50
0.40
0.30
0.20
0.10
0.00
G.gallus-S.scrofa
G.gallus-H.sapiens
H.sapiens-S.scrofa
c RSCU distances for avian virus–host pairs
2.50
2.00
1.50
1.00
0.50
0.00
Avian-G.gallus
Avian-S.scrofa
Avian-H.sapiens
d RSCU distances for swine virus–host pairs
3.50
3.00
2.50
2.00
1.50
1.00
0.50
0.00
Swine-S.scrofa
Swine-G.gallus
Swine-H.sapiens
e RSCU distances for human virus–host pairs
3.50
3.00
2.50
2.00
1.50
1.00
0.50
0.00
Human-H.sapiens
Human-G.gallus
Human-S.scrofa

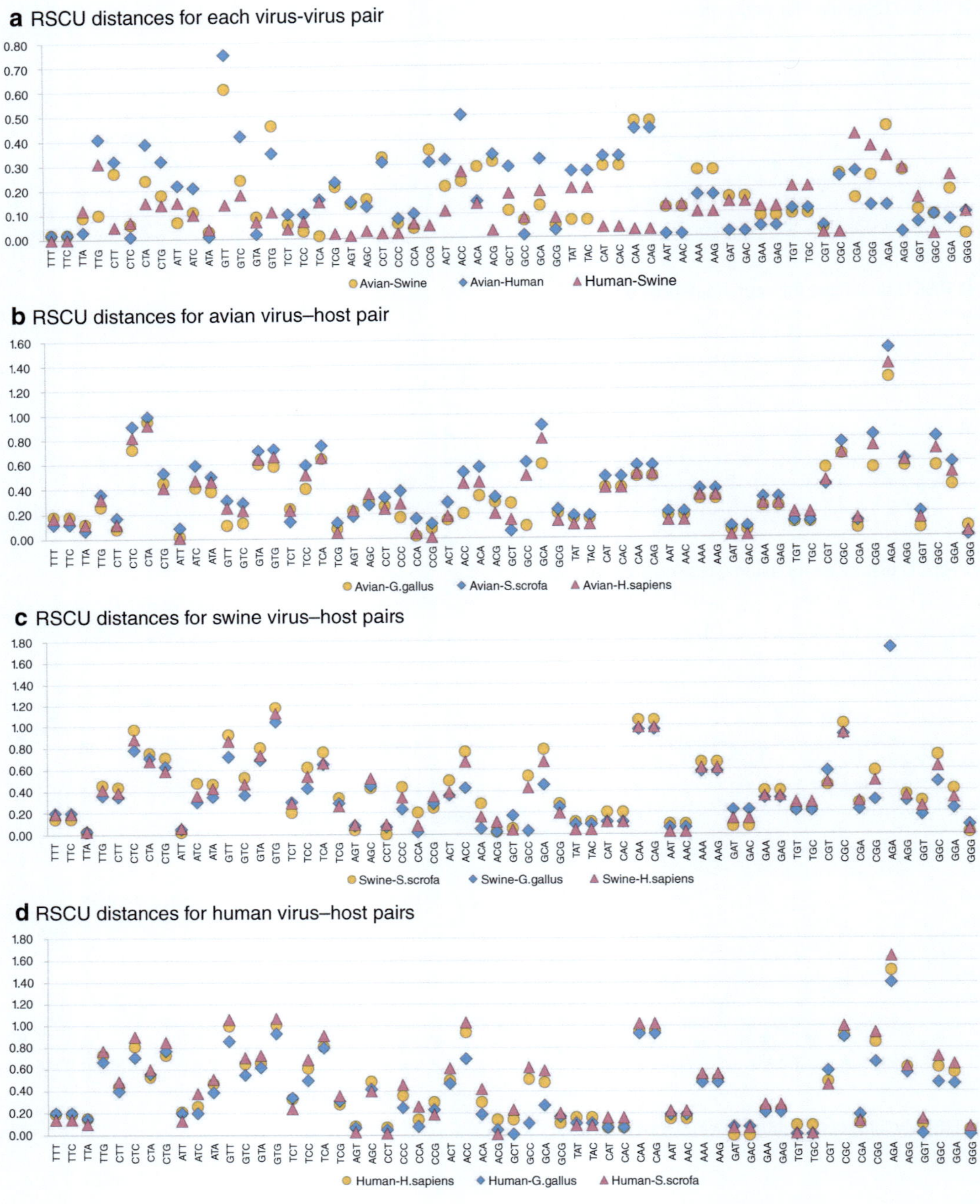

Fig. 4.4 Pairwise RSCU distance profiles between viruses and hosts in the H3N2 subtype. This figure illustrates the pairwise comparisons of relative synonymous codon usage (RSCU) distances in the HA-coding regions of H3N2 influenza A viruses and their corresponding host genomes. The panels present RSCU distance patterns for different types of pairings: (**a**) comparisons between different H3N2 virus groups (avian, swine, and human); (**b**) RSCU distances between avian H3N2 viruses and each host genome; (**c**) RSCU distances between swine H3N2 viruses and each host genome; (**d**) RSCU distances between human H3N2 viruses and each host genome. Each data point represents the codon-specific RSCU distance for a given pairwise comparison, providing insight into codon usage divergence across virus–virus and virus–host relationships

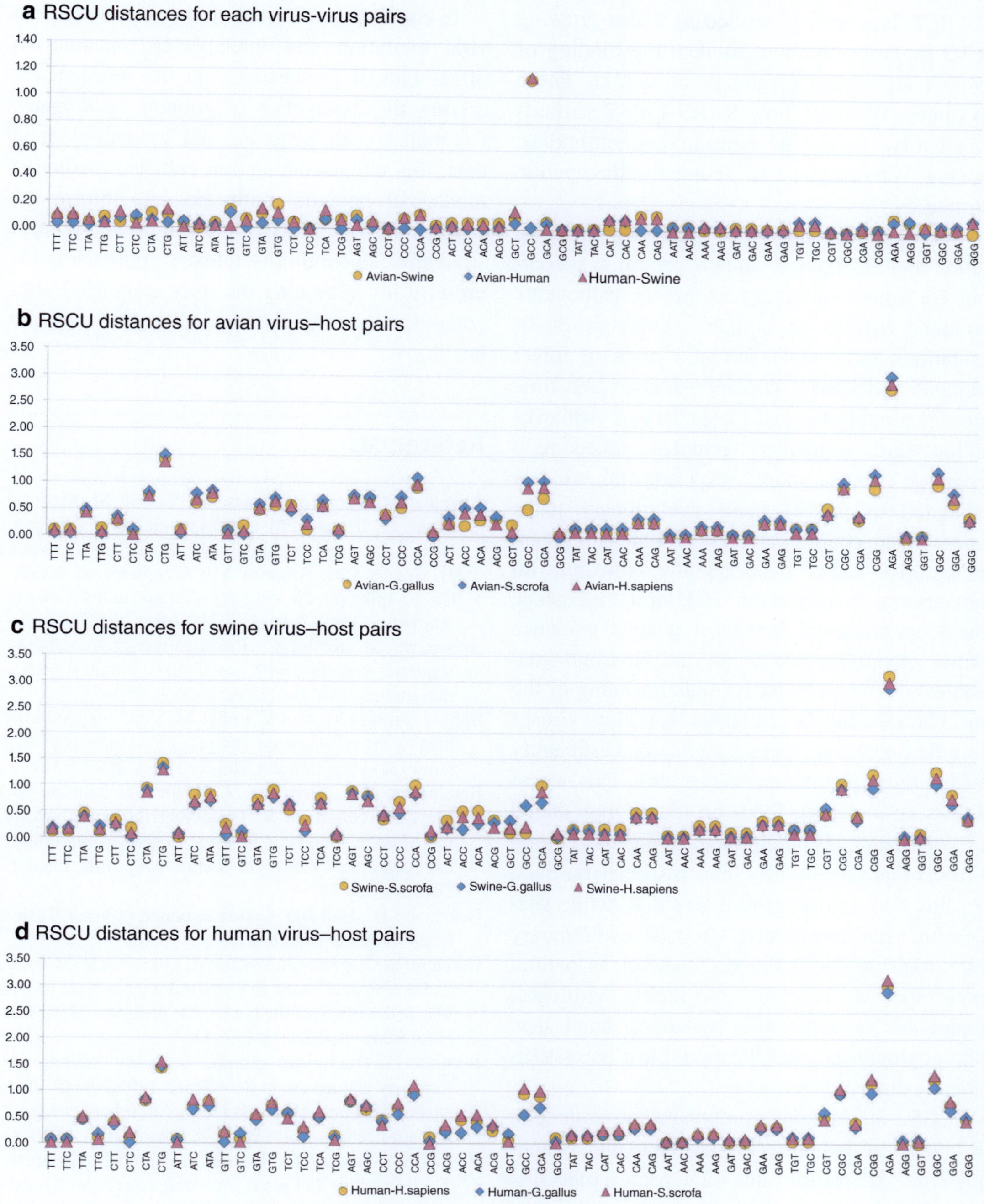

Fig. 4.5 Pairwise RSCU distance profiles between viruses and hosts in the H5N1 subtype. This figure illustrates the pairwise comparisons of relative synonymous codon usage (RSCU) distances in the HA-coding regions of H5N1 influenza A viruses and their corresponding host genomes. The panels present RSCU distance patterns for different types of pairings: (**a**) comparisons between different H5N1 virus groups (avian, swine, and human); (**b**) RSCU distances between avian H5N1 viruses and each host genome; (**c**) RSCU distances between swine H5N1 viruses and each host genome; (**d**) RSCU distances between human H5N1 viruses and each host genome. Each data point represents the codon-specific RSCU distance for a given pairwise comparison, providing insight into codon usage divergence across virus–virus and virus–host relationships

B.1.427, has been identified as a fast-growing RBD mutation that can weaken the binding of known antibodies (Wang et al. 2021). These examples illustrate how SARS-CoV-2 variants can evolve to escape neutralizing antibodies, posing challenges to current therapeutic strategies.

The ongoing challenge of predicting viral evolution and emergence remains central to preparing for future outbreaks of highly pathogenic zoonotic viruses in humans. Although minor mutations can enable animal viruses to infect humans and sustain human-to-human transmission, predicting the exact trajectory of viral evolution and adaptation remains exceedingly complex. Furthermore, experimental studies designed to predict these evolutionary paths under ideal conditions may present significant biosecurity risks. Consequently, research on interspecies transmission and viral emergence must be grounded in robust genetic evidence while accounting for social and environmental contexts. A comprehensive understanding of the mechanisms and determinants that allow viruses to infect new host species and spread efficiently within those populations is essential. This necessitates an increased focus on identifying shared evolutionary and epidemiological patterns among viruses that successfully breach species barriers. To this end, critical parameters that define successful viral emergence, such as evolutionary rate, transmission pathways, reservoir or natural host species, receptor specificity, virulence, pathogenicity, replication dynamics, burst size, and phylogenetic classification, must be systematically characterized.

CUB, an integral component of molecular evolution, adds another dimension to understanding viral adaptation and interspecies transmission. This bias, reflecting the preferential use of certain codons in protein encoding, is influenced by factors such as tRNA availability, GC content, and selective pressures. CUB can significantly impact viral replication efficiency, protein expression, and host immune evasion. Investigating CUB in viruses provides valuable insights into how they adapt to new hosts and optimize their replication machinery.

In conclusion, elucidating the role of CUB in viral evolution and interspecies transmission offers critical perspectives on the mechanisms driving the emergence of zoonotic pathogens. Combined with genomic and epidemiological data, this understanding can enhance predictive models of potential outbreaks and inform the development of effective prevention and control strategies. This multidisciplinary approach holds promise for mitigating the risks associated with zoonotic viruses and safeguarding global public health.

References

Alenquer M, Ferreira F, Lousa D, Valério M, Medina-Lopes M, Bergman M-L, Gonçalves J, Demengeot J, Leite RB, Lilue J, Ning Z, Penha-Gonçalves C, Soares H, Soares CM, Amorim MJ. Signatures in SARS-CoV-2 spike protein conferring escape to neutralizing antibodies. PLoS Pathog. 2021;17(8):e1009772.

Anisimova M, Kosiol C. Investigating protein-coding sequence evolution with probabilistic codon substitution models. Mol Biol Evol. 2009;26:255–71.

Bahir I, Fromer M, Prat Y, Linial M. Viral adaptation to host: a proteome-based analysis of codon usage and amino acid preferences. Mol Syst Biol. 2009;5:311.

Barahimipour R, Strenkert D, Neupert J, Schroda M, Merchant SS, Bock R. Dissecting the contributions of GC content and codon usage to gene expression in the model alga *Chlamydomonas reinhardtii*. Plant J. 2015;84:704–17.

Bennetzen JL, Hall BD. Codon selection in yeast. J Biol Chem. 1982;257:3026–31.

Berkhout B, Grigoriev A, Bakker M, Lukashov VV. Codon and amino acid usage in retroviral genomes is consistent with virus-specific nucleotide pressure. AIDS Res Hum Retrovir. 2002;18:133–41.

Bernardi G. The human genome: organization and evolutionary history. Annu Rev Genet. 1995;29:445–76.

Bisen PS, Debnath M, Prasad GB. Microbes: concepts and applications. John Wiley & Sons; 2012.

Bussey KA, Bousse TL, Desmet EA, Kim B, Takimoto T. PB2 residue 271 plays a key role in enhanced polymerase activity of influenza A viruses in mammalian host cells. J Virol. 2010;84:4395–406.

Cannarozzi GM, Schneider A, editors. Codon evolution: mechanisms and models. Oxford Univ. Press; 2012.

Carbone A. Codon bias is a major factor explaining phage evolution in translationally biased hosts. J Mol Evol. 2008;66:210–23.

Castrucci MR, Donatelli I, Sidoli L, Barigazzi G, Kawaoka Y, Webster RG. Genetic reassortment between avian and human influenza A viruses in Italian pigs. Virology. 1993;193:503–6.

Chaney JL, Clark PL. Roles for synonymous codon usage in protein biogenesis. Annu Rev Biophys. 2015;44:143–66.

Cho M. Prediction study on the cross-species infection of viruses using bioinformatics methods. PhD Thesis, Graduate School of Public Health, Seoul National University, Republic of Korea; 2019.

Claas E, Kawaoka Y, de Jong J, Masurel N, Webster R. Infection of children with avian-human reassortant influenza virus from pigs in Europe. Virology. 1994;204:453–7.

Claas EC, Osterhaus AD, Van Beek R, De Jong JC, Rimmelzwaan GF, Senne DA, Krauss S, Shortridge K, Webster RG. Human influenza A H5N1 virus related to a highly pathogenic avian influenza virus. Lancet. 1998;351:472–7.

Coleman JR, Papamichail D, Skiena S, Futcher B, Wimmer E, Mueller S. Virus attenuation by genome-scale changes in codon pair bias. Science. 2008;320:1784–7.

Connor RJ, Kawaoka Y, Webster RG, Paulson JC. Receptor specificity in human, avian, and equine H2 and H3 influenza virus isolates. Virology. 1994;205:17–23.

Costantini M, Clay O, Auletta F, Bernardi G. An isochore map of human chromosomes. Genome Res. 2006;16:536–41.

Costantini M, Clay O, Federico C, Saccone S, Auletta F, Bernardi G. Human chromosomal bands: nested structure, high-definition map and molecular basis. Chromosoma. 2007;116:29–40.

Crick F, Barnett L, Brenner S, Watts-Tobin RJ. General nature of the genetic code for proteins. Nature. 1961;192:1227–32.

Ellegren H, Smith NG, Webster MT. Mutation rate variation in the mammalian genome. Curr Opin Genet Dev. 2003;13:562–8.

Eyre-Walker A, Hurst LD. The evolution of isochores. Nat Rev Genet. 2001;2:549.

Freire-Picos MA, Gonzalez-Siso MI, Rodríguez-Belmonte E, Rodríguez-Torres AM, Ramil E, Cerdan ME. Codon usage in *Kluyveromyces lactis* and in yeast cytochrome c-encoding genes. Gene. 1994;139:43–9.

de Graaf M, Fouchier RA. Role of receptor binding specificity in influenza A virus transmission and pathogenesis. EMBO J. 2014;33:823–41.

Grantham R, Gautier C, Gouy M, Jacobzone M, Mercier R. Codon catalog usage is a genome strategy modulated for gene expressivity. Nucleic Acids Res. 1981;9:213.

Gustafsson C, Govindarajan S, Minshull J. Codon bias and heterologous protein expression. Trends Biotechnol. 2004;22:346–53.

Ikemura T. Correlation between the abundance of *Escherichia coli* transfer RNAs and the occurrence of the respective codons in its protein genes: a proposal for a synonymous codon choice that is optimal for the *E. coli* translational system. J Mol Biol. 1981;151:389–409.

Ito T, Couceiro JNS, Kelm S, Baum LG, Krauss S, Castrucci MR, et al. Molecular basis for the generation in pigs of influenza A viruses with pandemic potential. J Virol. 1998;72:7367–73.

Jenkins GM, Holmes EC. The extent of codon usage bias in human RNA viruses and its evolutionary origin. Virus Res. 2003;92:1–7.

Kane JF. Effects of rare codon clusters on high-level expression of heterologous proteins in *Escherichia coli*. Curr Opin Biotechnol. 1995;6:494–500.

Kawaoka Y, Krauss S, Webster RG. Avian-to-human transmission of the PB1 gene of influenza A viruses in the 1957 and 1968 pandemics. J Virol. 1989;63:4603–8.

Knight RD, Freeland SJ, Landweber LF. A simple model based on mutation and selection explains trends in codon and amino-acid usage and GC composition within and across genomes. Genome Biol. 2001;2:research0010-1.

Komar AA. The Yin and Yang of codon usage. Hum Mol Genet. 2016;25:R77–85.

Kramvis A, Kew MC. Epidemiology of hepatitis B virus in Africa, its genotypes and clinical associations of genotypes. Hepatol Res. 2007;37:S9–S19.

Li F, Li W, Farzan M, Harrison SC. Structure of SARS coronavirus spike receptor-binding domain complexed with receptor. Science. 2005;309:1864–8.

Liu YS, Zhou JH, Chen HT, Ma LN, Pejsak Z, Ding YZ, Zhang J. The characteristics of the synonymous codon usage in enterovirus 71 virus and the effects of host on the virus in codon usage pattern. Infect Genet Evol. 2011;11:1168–73.

Lobo FP, Mota BE, Pena SD, Azevedo V, Macedo AM, Tauch A, Machado CR, Franco GR. Virus-host coevolution: common patterns of nucleotide motif usage in Flaviviridae and their hosts. PLoS One. 2009;4:e6282.

Lobry JR, Sueoka N. Asymmetric directional mutation pressures in bacteria. Genome Biol. 2002;3:1–14.

Mani I, Singh V, Chaudhary DK, Somvanshi P, Negi MPS. Codon optimization of the major antigen encoding genes of diverse strains of influenza A virus. Interdiscip Sci Comput Life Sci. 2011;3:36–42.

Masurel N, Marine WM. Recycling of Asian and Hong Kong influenza A virus hemagglutinins in man. Am J Epidemiol. 1973;97:44–9.

Mueller S, Papamichail D, Coleman JR, Skiena S, Wimmer E. Reduction of the rate of poliovirus protein synthesis through large-scale codon deoptimization causes attenuation of viral virulence by lowering specific infectivity. J Virol. 2006;80:9687–96.

Robinson M, Lilley R, Little S, Emtage JS, Yarranton G, Stephens P, Millican A, Eaton M, Humphreys G. Codon usage can affect efficiency of translation of genes in *Escherichia coli*. Nucleic Acids Res. 1984;12:6663–71.

Rocha EP. Codon usage bias from tRNA's point of view: redundancy, specialization, and efficient decoding for translation optimization. Genome Res. 2004;14:2279–86.

Roth A, Anisimova M, Cannarozzi GM. Measuring codon usage bias. In: Codon evolution: mechanisms

and models. Oxford University Press Inc.; 2012. p. 189–217.
Scholtissek C. Pigs as 'mixing vessels' for the creation of new pandemic influenza A viruses. Med Princ Pract. 1990;2:65–71.
Shackelton LA, Parrish RC, Holmes EC. Evolutionary basis of codon usage and nucleotide composition bias in vertebrate DNA viruses. J Mol Evol. 2006;62:551–63.
Sharp PM, Hahn BH. Origins of HIV and the AIDS pandemic. Cold Spring Harb Perspect Med. 2011;1:a006841.
Sharp PM, Li WH. An evolutionary perspective on synonymous codon usage in unicellular organisms. J Mol Evol. 1986;24:28–38.
Sharp PM, Li WH. The codon adaptation index—a measure of directional synonymous codon usage bias, and its potential applications. Nucleic Acids Res. 1987;15:1281–95.
Sharp PM, Stenico M, Peden JF, Lloyd AT. Codon usage: mutational bias, translational selection, or both? J Mol Evol. 1993;37:1–17.
Spencer PS, Barral JM. Genetic code redundancy and its influence on the encoded polypeptides. Comput Struct Biotechnol J. 2012;1:e201204006.
Subbarao EK, London W, Murphy BR. A single amino acid in the PB2 gene of influenza A virus is a determinant of host range. J Virol. 1993;67:1761–4.
Sueoka N. Variation and heterogeneity of base composition of deoxyribonucleic acids: a compilation of old and new data. J Mol Biol. 1961;3:31–IN15.
Sueoka N. On the genetic basis of variation and heterogeneity of DNA base composition. Proc Natl Acad Sci. 1962;48:582–92.
Tang XC, Zhang JX, Zhang SY, Wang P, Fan XH, Li LF, Li G, Dong BQ, Liu W, Cheung CL, Xu KM, Song WJ, Vijaykrishna D, Poon LL, Peiris JS, Smith GJ, Chen H, Guan Y. Prevalence and genetic diversity of coronaviruses in bats from China. J Virol. 2006;80:7481–90.
Taubenberger JK, Reid AH, Lourens RM, Wang R, Jin G, Fanning TG. Characterization of the 1918 influenza virus polymerase genes. Nature. 2005;437:889–93.
Vavouri T, Lehner B. Chromatin organization in sperm may be the major functional consequence of base composition variation in the human genome. PLoS Genet. 2011;7:e1002036.
Wan XF, Xu D, Kleinhofs A, Zhou J. Quantitative relationship between synonymous codon usage bias and GC composition across unicellular genomes. BMC Evol Biol. 2004;4:19.
Wang R, Chen J, Gao K, Wei GW. Vaccine-escape and fast-growing mutations in the United Kingdom, the United States, Singapore, Spain, India, and other COVID-19-devastated countries. Genomics. 2021;113:2158–70.
Watson JD, Crick FHC. A structure for deoxyribose nucleic acid. Nature. 1953;171:737–8.
Weaver SC, Barrett AD. Transmission cycles, host range, evolution and emergence of arboviral disease. Nat Rev Microbiol. 2004;2:789–801.
Webster RG, Bean WJ, Gorman OT, Chambers TM, Kawaoka Y. Evolution and ecology of influenza A viruses. Microbiol Rev. 1992;56:152–79.
Wong EH, Smith DK, Rabadan R, Peiris M, Poon LL. Codon usage bias and the evolution of influenza A viruses. BMC Evol Biol. 2010;10:253.
Wright F. The 'effective number of codons' used in a gene. Gene. 1990;87:23–9.
Zhou J, Liu WJ, Peng SW, Sun XY, Frazer I. Papillomavirus capsid protein expression level depends on the match between codon usage and tRNA availability. J Virol. 1999;73:4972–82.
Zoubak S, Clay O, Bernardi G. The gene distribution of the human genome. Gene. 1996;174:95–102.

5 Artificial Intelligence Algorithms

Artificial Intelligence (AI) has become a transformative force across various fields, offering innovative solutions to complex challenges with remarkable precision and efficiency. At its core, AI relies on algorithms—systematic procedures for data processing, automated reasoning, and decision-making. These algorithms are fundamental to AI's versatility, able to harness vast data sets and computational power to derive meaningful insights and guide impactful decisions.

AI algorithms can be classified based on learning paradigms, structural architectures, combinatorial methodologies, and task-specific adaptability. Further classifications include supervised, unsupervised, and reinforcement learning, as well as hybrid methods like ensemble learning and advanced neural networks. Specialized techniques such as probabilistic reasoning, dimensionality reduction, and uncertainty handling further broaden AI's ability to tackle highly specialized and domain-specific problems. A solid understanding of these categories is essential for selecting the most effective algorithm to optimize performance and efficiency.

This chapter begins by categorizing different AI algorithms and providing a detailed exploration of how they are tailored to address specific data types and real-world applications. This includes the handling of structured and unstructured data, as well as the integration of hybrid approaches for enhanced accuracy. The chapter then shifts focus towards the transformative applications of AI in virology and related fields. For example, AI-powered systems are revolutionizing virus classification, accelerating antiviral drug and vaccine development, and improving diagnostic precision. Predictive modeling now allows epidemiologists to forecast disease outbreaks and refine public health strategies. Finally, we present a case study on the application of machine learning techniques for the analysis of protein determinants associated with host-specific infection mechanisms in polyomaviruses (PyVs). This example demonstrates the practical integration of AI algorithms in virology, showing their use for virus classification, host-virus interaction prediction, and the explanation of intricate biological processes. The insights derived from this study highlight the potential of machine learning to deepen our understanding of viral dynamics, host system regulation, and disease progression.

Through this structured approach, this chapter aims to deliver a comprehensive exploration of AI algorithms and their varied applications. By examining foundational principles, real-world implementations, and a detailed case study, we emphasize the transformative potential of AI for addressing complex scientific challenges, advancing research endeavors, and driving innovation across diverse fields, with a focus on the realm of virology.

H. S. Son, M. Cho, *Viroinformatics and Computational Virology in Public Health*,
https://doi.org/10.1007/978-981-96-5757-5_5

5.1 Comprehensive Categorization of AI Algorithms

AI algorithms are meticulously designed to address specific tasks and adapt to the nature of the data they process, exploiting their versatility and relevance across a broad range of applications. These algorithms are generally classified based on their learning approaches, structural architectures, the type of data they handle, and their specialization in solving complex computational problems.

Some algorithms work with labeled datasets, meaning the relationships between inputs and outputs are explicitly defined. This method is known as *supervised learning*. Algorithms such as linear regression, logistic regression, support vector machines (SVM), and neural networks fall into this category. These approaches excel in tasks such as classification, where outcomes are categorized, or regression, where predictions are continuous. In contrast, other algorithms function without labeled data, by discovering hidden patterns or structures. These *unsupervised learning* algorithms, including k-means clustering, principal component analysis (PCA), and autoencoders, are crucial for clustering, anomaly detection, and dimensionality reduction. Bridging these two approaches, *semi-supervised learning* algorithms integrate labeled and unlabeled data to improve model performance. Techniques like self-training and co-training are particularly useful in situations where labeled data is limited, leveraging often-abundant unlabeled data to enhance learning outcomes. Meanwhile, *reinforcement learning* enables agents to interact with their environment and develop optimal actions through feedback in the form of rewards or penalties. Algorithms such as Q-learning, deep Q-networks (DQN), and actor-critic models have widespread applications in robotics, autonomous systems, and gaming.

The structural design of an AI model plays a significant role in its functionality. Traditional models, such as ridge regression, decision trees, and random forest (RF), are rooted in statistical methods and classical machine learning techniques. They are simple, interpretable, and highly effective for structured data. *Neural networks*, inspired by the interconnected structure of the human brain, are adept at capturing complex and nonlinear relationships within data and form the basis for deep learning. *Deep learning* models, which extend neural networks into multi-layered architectures, have driven significant advances in areas like image recognition, natural language processing, and speech analysis. Key examples include convolutional neural networks (CNNs) for image data, recurrent neural networks (RNNs) for sequential data, long short-term memory (LSTM) networks for time-series tasks, and transformers for the understanding and generation of natural language.

To enhance predictive accuracy and robustness, AI algorithms often utilize *ensemble learning* techniques. While individual algorithms, such as decision trees and k-nearest neighbors (kNN), operate independently to address specific tasks, ensemble methods—which include RF, bagging, boosting, and stacking—combine multiple models to improve performance and reduce errors, ensuring reliability.

The type of data an algorithm processes significantly influences its design and functionality. *Structured data*, which is organized and relationally consistent, is best handled by models like logistic regression and SVM, which are well-suited for managing well-defined variable relationships. On the other hand, *unstructured data*—such as text, images, or videos—requires more advanced approaches, such as *neural networks* and *deep learning architectures*, which include LSTM networks and generative adversarial networks (GANs). These algorithms excel at processing and interpreting unstructured data, driving innovations in image generation, speech recognition, and text translation.

In addition to these broad categories, certain algorithms are specialized for addressing specific computational challenges. *Dimensionality reduction* techniques like PCA and t-distributed stochastic neighbor embedding (t-SNE) simplify high-dimensional datasets by retaining key information while reducing complexity. *Fuzzy logic systems*, such as fuzzy C-means clustering and fuzzy inference systems, excel in managing uncertainty and imprecision, making decisions in environments where accuracy is difficult. *Probabilistic algorithms*, such as Naïve Bayes (NB) and Bayesian networks, utilize Bayes' theorem for reasoning under uncertainty; these are widely applied in classification and prediction tasks. *Simulation-based approaches*, including SARSA (state-action-reward-state-action) and ant colony optimization, are particularly effective for modeling real-world scenarios and navigating complex decision-making paths.

Together, these diverse algorithms form the backbone of contemporary AI, each being designed for specific tasks and data types. Their adaptability and sophistication empower them to drive innovation, solve complex challenges, and improve efficiency across a wide range of industries. This chapter focuses on key algorithms from selected categories. For additional information on other AI algorithms, please refer to supplementary publications.

5.1.1 Linear Regression

Origin The least squares method, a foundational principle of linear regression, was clearly formulated and published by Adrien-Marie Legendre in 1805. Carl Friedrich Gauss formally described its application in 1809 in his work on calculating celestial orbits, extending the method by connecting it to probability theory and the normal distribution. His systematic approach to minimizing the sum of squared deviations between observed and predicted values laid the groundwork for regression analysis, with applications extending far beyond its astronomical origins to a wide array of scientific and practical domains (Gauss 1809).

Purpose Linear regression is used to predict a continuous output based on the linear relationship between the input variables (features) and the output variable.

Mechanism The algorithm fits a linear equation to the observed data. The simplest form is the univariate linear regression, where the relationship is between a single feature and the target variable. For multiple features, it is referred to as multiple linear regression.

Mathematical Formula

$$Y = \beta_0 + \beta_1 X_1 + \beta_2 X_2 + \ldots + \beta_n X_n + \epsilon$$

where Y is the dependent variable, $X_1, X_2, \ldots, X_n$ are the independent variables, β_0 is the y-intercept, $\beta_1, \beta_2, \ldots, \beta_n$ are the coefficients, and ϵ is the error term.

Training Process The model parameters (coefficients) are estimated using the least squares method, which minimizes the sum of the squared differences between the observed and predicted values.

Application Linear regression has been applied to model the relationship between COVID-19 transmission rates and environmental factors such as temperature and humidity, yielding insights into the seasonal dynamics of the virus (Ma et al. 2020).

5.1.2 Lasso Regression

Origin Lasso (Least Absolute Shrinkage and Selection Operator) regression was introduced by Robert Tibshirani in 1996. Tibshirani's work addressed the challenge of selecting relevant predictors in high-dimensional data, combining regularization with variable selection to enhance interpretability and accuracy in predictive models (Tibshirani 1996).

Purpose Lasso regression is used for both regularization and variable selection. It helps improve the prediction accuracy and interpretability of a statistical model.

Mechanism Lasso regression adds an L1 regularization term to the cost function, which penalizes the absolute values of the coefficients.

Mathematical Formula

$$Minimize \sum_{n}^{i=1} \left(Y_i - \widehat{Y}_i\right)^2 + \lambda \sum_{p}^{j=1} \left|\beta_j\right|$$

where λ controls the degree of regularisation.

Training Process Lasso uses coordinate descent or iterative shrinkage methods to minimize the loss function. The L1 penalty can shrink some coefficients to zero, thus selecting variables and making the model more interpretable.

Application Lasso regression has been utilized to identify key genetic markers associated with the antigenicity of influenza A (H1N1) viruses by selecting significant features, such as mutations and variations in glycosylation of hemagglutinin (HA), from extensive genomic datasets (Li et al. 2020).

5.1.3 Elastic Net Regression

Origin Proposed by Hui Zou and Trevor Hastie in 2005, Elastic Net regression addresses limitations in Lasso regression when predictors are highly correlated. By introducing a combination of L1 (Lasso) and L2 (Ridge) penalties, Elastic Net enhances model stability and prediction accuracy in the presence of multicollinearity (Zou and Hastie 2005).

Purpose Elastic Net regression combines the penalties of Ridge and Lasso regressions to improve model performance when dealing with highly correlated predictors.

Mechanism The algorithm introduces a mix of L1 and L2 regularization terms to the cost function.

Mathematical Formula

$$Minimize \sum_{n}^{i=1} \left(Y_i - \widehat{Y}_i\right)^2 + \lambda \sum_{p}^{j=1} \left|\beta_j\right| + \lambda_2 \sum_{p}^{j=1} \beta_j^2$$

where λ_1 is the L1 regularisation weight and λ_2 is the L2 regularization weight.

Training Process Elastic Net solves a combined L1–L2 penalized objective function using coordinate descent, iteratively optimizing the weights to effectively handle correlated predictors.

Application Elastic Net regression has been used to investigate the relationship between Lyme disease incidence and environmental as well as socio-economic factors. A spatial Poisson regression model, combined with an adaptive Elastic Net penalty, has been developed to effectively identify significant variables influencing Lyme disease occurrence in Virginia, while handling correlated predictors in complex datasets (Xie et al. 2019).

5.1.4 Logistic Regression

Origin The logistic function was first introduced by Pierre François Verhulst in 1838 to model population growth. The application of logistic regression to binary classification was subsequently formalized by David Cox in 1958, establishing it as a key statistical method for analyzing binary outcomes (Cox 1958).

Purpose Logistic regression is used to predict the probability of a categorical dependent variable. It is primarily used for binary classification but can be extended to multi-class classification.

Mechanism This algorithm models the relationship between a set of input features and a binary outcome using the logistic function. The logistic function, or sigmoid function, maps any real-valued number into the range [0, 1], so it can then be interpreted as a probability.

Mathematical Formula

$$P\left(y = 1 \mid X\right) = \frac{1}{1 + e^{-(\beta_0 + \beta_1 X_1 + \beta_2 X_2 + \ldots + \beta_n X_n)}}$$

Training Process The model parameters (coefficients) are estimated using maximum likelihood estimation, which seeks to maximize the probability of observing the given dataset.

Application Logistic regression has been used to analyze factors such as age, comorbidities, and inflammatory markers to predict severe outcomes in COVID-19 patients, aiding in clinical decision-making (Wynants et al. 2020).

5.1.5 Support Vector Machine (SVM)

Origin The concept of SVMs was introduced by Vladimir Vapnik and Alexey Chervonenkis in 1963. They proposed a robust mathematical framework to classify data by identifying hyperplanes that maximize the margin between different classes. This paved the way for modern machine learning classification techniques (Vapnik and Chervonenkis 1964).

Purpose SVM is used to classify data by finding the hyperplane that best separates the data into different classes. It is effective for high-dimensional spaces and cases where the number of dimensions exceeds the number of samples.

Mechanism SVM constructs a hyperplane or set of hyperplanes in a high-dimensional space. The objective is to find a hyperplane that maximizes the margin between the classes. Support vectors are the data points that are closest to the hyperplane and influence its position and orientation.

Mathematical Formula

$$Maximize \frac{2}{\omega}$$

subject to the constraint that each point must be correctly classified by the margin.

$$y_i\left(\omega^T x_i + b\right) \geq 1$$

where ω is the hyperplane normal vector, and b is the offset.

Training Process SVM solves a quadratic optimization problem to maximize the margin between classes. For nonlinear data, the kernel trick maps data into a higher-dimensional space. Support vectors determine the hyperplane.

Application SVMs have been employed to classify dengue fever patients based on gene expression data, enabling differentiation between various clinical presentations, such as dengue fever and dengue hemorrhagic fever, supporting more accurate diagnostics and targeted treatment strategies (Gomes et al. 2010).

5.1.6 K-Nearest Neighbors (kNN)

Origin kNN was introduced by Evelyn Fix and Joseph Hodges in 1951 as a nonparametric method for discriminant analysis. Their pioneering work laid the foundation for proximity-based classification and regression algorithms, particularly suited for intuitive data analysis in low-dimensional spaces (Fix and Hodges 1951).

Purpose kNN is a nonparametric algorithm used for classification by identifying the training examples in the feature space that are closest to the input data point.

Mechanism The algorithm stores all available cases and classifies new cases based on a similarity measure (e.g., distance functions). For a given data point, kNN identifies the k-nearest neighbors and assigns the classification that is most frequent among them.

Mathematical Formula (Distance Metrics)
Euclidean Distance:

$$d\left(x,y\right) = \sqrt{\sum_{n}^{i=1}\left(x_i - y_i\right)^2}$$

Manhattan Distance:

$$d\left(x,y\right) = \sum_{n}^{i=1}\left|x_i - y_i\right|$$

Minkowski Distance:

$$d(x,y) = \left(\sum_{n}^{i=1} |x_i - y_i|^p \right)^{1/p}$$

Training Process kNN stores the entire training dataset. For each query point, it calculates distances to all training points and assigns the majority class among the k-nearest neighbors.

Application kNN has been applied to distinguish SARS-CoV-2 from common human coronaviruses, such as AlphaCoV, BetaCoV-1, MERS-CoV, HKU1-CoV, NL63-CoV, and 229E-CoV, using CpG island features, demonstrating its effectiveness as a classifier for genomic data analysis (Arslan and Arslan 2021).

5.1.7 K-Means Clustering

Origin Stuart Lloyd introduced the K-Means algorithm in 1957 while working on signal quantization problems at Bell Labs. His work introduced a method to iteratively partition datasets into clusters by minimizing the sum of squared distances between data points and their respective cluster centroids (Lloyd 1982).

Purpose K-Means clustering aims to partition data into k distinct clusters based on feature similarity.

Mechanism The algorithm iteratively assigns each data point to one of k clusters so that the sum of the squared distances from each point to the centroid of its assigned cluster is minimized.

Mathematical Formula

$$Minimize \sum_{k}^{i=1} \sum_{x \in C_i} \|x - \mu_i\|^2$$

where C_i is the set of points in cluster i, and μ_i is the centroid of cluster i.

Training Process

- Initialization: Choose k initial centroids randomly.
- Assignment: Assign each data point to the nearest centroid.
- Update: Recalculate the centroids as the mean of all points assigned to that cluster.
- Iteration: Repeat the assignment and update steps until centroids stabilize.

Application K-Means clustering has been employed to categorize provinces in Indonesia based on COVID-19 risk levels, providing insights into regional pandemic dynamics and enabling targeted public health interventions (Abdullah et al. 2022).

5.1.8 Hierarchical Clustering

Origin The concept of hierarchical clustering was formalized by S. C. Johnson in 1967. Johnson's methodology introduced agglomerative and divisive clustering approaches, enabling the creation of nested clusters visualized through dendrograms (Johnson 1967).

Purpose Hierarchical clustering builds a nested hierarchy of clusters, either in an agglomerative (bottom-up) or divisive (top-down) manner.

Mechanism The agglomerative approach starts with each data point as a single cluster and merges the closest pairs of clusters iteratively until all points belong to one cluster. Conversely, divisive clustering starts with one cluster and recursively splits it.

Mathematical Formula

Agglomerative: Minimizes the distance between clusters, often using methods like single-linkage (minimum distance), complete-linkage (maximum distance), or average-linkage.

$$d(C_i, C_i) = \min_{x \in C_i,\, y \in C_j} \|x - y\|$$

Training Process The algorithm constructs a dendrogram (tree) that visualizes the order of cluster merges or splits. The process continues until a single cluster remains or the desired number of clusters is achieved.

Application Hierarchical clustering has been used to construct phylogenetic trees of SARS-CoV-2 variants, revealing evolutionary relationships and structural characteristics of the spike protein, including its proteolytic activation loop (Jaimes et al. 2020).

5.1.9 Decision Trees

Origin The Iterative Dichotomiser 3 (ID3) algorithm, developed by J. Ross Quinlan in 1986, marked a significant milestone in decision tree research. Quinlan's work introduced an efficient approach for data partitioning using criteria such as information gain, revolutionizing classification tasks (Quinlan 1986).

Purpose Decision trees are used to classify data by learning simple decision rules inferred from the features of the data. They are popular for their interpretability and ease of understanding.

Mechanism The algorithm splits the dataset into subsets based on the value of input features. This splitting is done recursively to form a tree structure, where each internal node represents a "test" of an attribute. Each branch then represents the outcome of the test, and each leaf node represents a class label.

Splitting Criteria Common criteria include Gini impurity and information gain.

Gini Impurity:

$$\text{Gini} = 1 - \sum_{n}^{i=1} p_i^2$$

Information Gain:

$$\text{IG} = H(\text{parent}) - \left[\frac{N_{\text{left}}}{N} H(\text{left}) + \frac{N_{\text{right}}}{N} H(\text{right})\right]$$

Entropy:

$$H(S) = -\sum_{c}^{i=1} p_i \log(p_i)$$

Training Process The tree is built by selecting the best feature to split the data at each node, based on the chosen splitting criterion. The process continues until a stopping criterion is met (e.g., all nodes are pure or a maximum depth is reached).

Application Decision trees based on clinical features have been used to predict the likelihood of seasonal influenza in patients presenting with influenza-like symptoms, helping clinicians to stratify patient risk during outbreaks (Afonso et al. 2012).

5.1.10 Random Forest

Origin RFs were introduced by Leo Breiman in 2001 as an ensemble learning method. Breiman's innovation combined multiple decision trees with bootstrap sampling and random feature selection, significantly enhancing prediction accuracy and robustness (Breiman 2001).

Purpose RFs aim to improve the accuracy and robustness of decision trees by combining the predictions of multiple trees.

Mechanism This algorithm constructs multiple decision trees during training; the mode of the classes (classification) or mean prediction (regression) of these individual trees are the outputs. Each tree is trained on a random subset of the data and features.

Mathematical Formula
For classification:

$$\hat{y} = \text{mode}\left(\{h_i(x)\}_{i=1}^{N}\right)$$

For regression:

$$\hat{y} = \frac{1}{N}\Sigma_{i=1}^{N} h_i(x)$$

where $h_i(x)$ is the prediction of the i-th tree and N is the total number of trees.

Training Process The training process involves creating multiple decision trees using bootstrap samples (random sampling with replacement) from the training dataset. Each tree is trained on a different subset of data, and a random subset of features is considered for splitting at each node. The final prediction is made by aggregating the predictions of all trees.

Application RFs have been employed to identify critical variables associated with bovine viral diarrhea virus infections, using robust machine learning techniques to analyze complex datasets and enhance disease prediction accuracy (Machado et al. 2015).

5.1.11 Bagging

Origin Bagging, short for Bootstrap Aggregating, was proposed by Leo Breiman in 1996. This method addressed the challenge of overfitting in predictive models by aggregating predictions from multiple resampled datasets, improving both stability and accuracy (Breiman 1996).

Purpose Bagging aims to reduce variance and prevent overfitting by combining the predictions of multiple models trained on different subsets of the data.

Mechanism Multiple models are trained on different bootstrap samples (random samples with replacement) of the training data. The final prediction is made by averaging (regression) or majority voting (classification) the predictions of all models.

Mathematical Formula
For classification:

$$\hat{y} = \text{mode}\left(\{h_i(x)\}_{i=1}^{M}\right)$$

For regression:

$$\hat{y} = \frac{1}{M}\sum_{i=1}^{M} h_i(x)$$

where $h_i(x)$ is the prediction of the i-th model and M is the total number of models.

Training Process The training process involves generating multiple bootstrap samples from the training dataset. A separate model is trained on each sample. The predictions from all models are combined to produce the final output, through either averaging or voting.

Application Bagging has been used to enhance the robustness of influenza diagnostic models by combining predictions from multiple machine learning algorithms, demonstrating enhanced accuracy and reliability across diverse datasets (Kumar et al. 2024).

5.1.12 Boosting

Origin AdaBoost, a key boosting algorithm, was introduced by Yoav Freund and Robert Schapire in 1995. Their work focused on sequentially combining weak learners to create a strong classifier, addressing errors from previous iterations (Freund and Schapire 1997).

Purpose Boosting improves model performance by training models sequentially, with each new model focusing on correcting the errors made by previous ones.

Mechanism Models are trained sequentially, with each new model giving more weight to data points that were misclassified by previous models. The final prediction is the weighted sum of the predictions from all models.

Mathematical Formula
For classification (AdaBoost):

$$F(x) = sign\left(\sum_{t=1}^{T} \alpha_t h_t(x)\right)$$

where α_t is the weight for model t, $h_t(x)$ is the t-th weak learner, and T is the total number of models.

Training Process The training process involves sequentially adding models. Each new model is trained to correct the errors of the combined ensemble of previous models. The model's predictions are weighted based on their accuracy, and these weights are used to update the importance of different data points in subsequent training.

Application Boosting algorithms, including eXtreme Gradient Boosting (XGB) and Light Gradient Boosting Machine (LGBM), have been applied to classify COVID-19 variants based on DNA sequence data, achieving improved accuracy in identifying variants and aiding pandemic response efforts (Muhammad et al. 2022).

5.1.13 Stacking

Origin David H. Wolpert introduced the concept of Stacked Generalization, or Stacking, in 1992. Wolpert's framework demonstrated the effectiveness of combining predictions from multiple base models using a meta-model to improve overall accuracy (Wolpert 1992).

Purpose Stacking aims to improve model performance by combining multiple models using a meta-model to make final predictions.

Mechanism Base models (level-0 models) are trained on training data, and a meta-model (level-1 model) is trained on the outputs of these base models. The final prediction is made by the meta-model.

Mathematical Formula

$$\hat{y} = h_{\text{meta}}\left(h_1(x), h_2(x), \ldots, h_N(x)\right)$$

where h_{meta} is the meta-model, $h_i(x)$ are the predictions of the base models, and N is the number of base models.

Training Process The training process involves first fitting multiple base models to the training data. The outputs of these base models are then used as input features to train a meta-model. The meta-model learns to combine the predictions of the base models to improve overall performance.

Application Stacking has been used to predict the incidence of infectious diarrhea by combining multiple prediction models. Using surveillance data on cases, symptoms, and meteorological factors, this approach has demonstrated improved predictive performance compared to individual models (Wang et al. 2024).

5.1.14 Naïve Bayes (NB)

Origin NB is based on Bayes' theorem, introduced by Thomas Bayes in the eighteenth century. Its simplistic assumption of feature independence has made it a cornerstone in probabilistic classification (Bayes 1763).

Purpose NB classifies data based on applying Bayes' theorem, with strong assumptions of independence between features.

Mechanism The posterior probability is calculated for each class based on the prior probability and the likelihood of the features, given the class.

Mathematical Formula

$$P\left(C_k \mid, x_1, x_2, \ldots, x_n\right) = \frac{P\left(C_k\right)\prod_{n}^{i=1} P\left(x_i \mid C_k\right)}{P\left(x_1, x_2, \ldots, x_n\right)}$$

where $P(C_k)$ is the prior probability of class C_k, and $P(x_i| C_k)$ is the likelihood of feature x_i given class C_k.

Training Process The process involves calculating the prior probability of each class and the likelihood of each feature, according to the class of the training data. These probabilities are then used to classify new instances by computing the posterior probability for each class and selecting the class with the highest probability.

Application NB has been applied to predict phage virion proteins using amino acid and dipeptide compositions, coupled with novel feature selection techniques, highlighting its utility in distinguishing viral components and its potential in phage proteomics research (Feng et al. 2013).

5.1.15 Principal Component Analysis (PCA)

Origin PCA was introduced by Karl Pearson in 1901 as a statistical method for reducing dimensionality while preserving variance. Pearson's work laid the foundation for modern multivariate data analysis (Pearson 1901). The method was later formalized and extended by Harold Hotelling in 1933, providing the mathematical framework commonly used today (Hotelling 1933).

Purpose PCA reduces the dimensionality of data while preserving as much variance as possible.

Mechanism PCA transforms the original data into a new coordinate system, where the directions of greatest variances are aligned with the first coordinates (principal components).

Mathematical Formula

$$Z = XW$$

where Z is the transformed data, X is the original data, and W is the matrix of eigenvectors of the covariance matrix of X.

Training Process The process involves standardizing the data, calculating the covariance matrix, extracting the eigenvalues and eigenvectors of the covariance matrix, and then projecting the original data onto the principal components.

Application PCA has been used to analyze large-scale genome sequences of SARS-CoV-2, providing rapid insights into viral evolution and mutation patterns while serving as a pre-processing step for advanced methods, such as decision tree-based analyses (Wang and Jiang 2021).

5.1.16 t-Distributed Stochastic Neighbor Embedding (t-SNE)

Origin t-SNE was developed by Laurens van der Maaten and Geoffrey Hinton in 2008 to visualize high-dimensional data by maintaining local similarities in a reduced-dimensional space (van der Maaten and Hinton 2008).

Purpose t-SNE reduces dimensions while maintaining the local structure of the data, making it suitable for visualizing high-dimensional data.

Mechanism High-dimensional Euclidean distances are converted into conditional probabilities that represent similarities. The Kullback-Leibler (KL) divergence between these probabilities is then minimized in high-dimensional and low-dimensional spaces.

Mathematical Formula

$$\mathrm{KL}\left(P \| Q\right) = \sum_{i \neq j} p_{ij} \log \frac{p_{ij}}{q_{ij}}$$

where p_{ij} is the probability distribution in the high-dimensional space, and q_{ij} is the probability distribution in the low-dimensional space.

Training Process The process starts by calculating the pairwise similarity probabilities in the high-dimensional space, then iteratively adjusts the points in the low-dimensional space to minimize the KL divergence between the two distributions.

Application t-SNE has been employed to analyze Zika virus proteomes, uncovering hidden patterns and visualizing clustering based on host and geographical factors. This dimensionality reduction technique has demonstrated its value in molecular epidemiology and evolutionary studies (Lara-Ramírez et al. 2024).

5.1.17 Neural Networks

Origin Neural networks were first conceptualized by Warren McCulloch and Walter Pitts in 1943. Their pioneering work on artificial neurons provided a computational model for simulating the interconnected structures of the brain (McCulloch and Pitts 1943). While their work laid the conceptual groundwork for modern artificial neural networks, the development of the field relied heavily on later advancements, such as the Perceptron in the 1950s (Rosenblatt 1958) and the Backpropagation algorithm in the 1980s (Rumelhart et al. 1986).

Purpose Neural networks are designed to recognize complex patterns and relationships in data. They are particularly effective for image recognition, natural language processing, and other tasks involving large datasets and intricate patterns.

Mechanism Neural networks consist of layers of interconnected nodes. Each node performs a weighted sum of its inputs and passes the result through an activation function. The network is trained by adjusting the weights to minimize the error between predicted and actual outputs.

Mathematical Formula
For a single neuron:

$$z = \sum_{i=1}^{n} \omega_i x_i + b$$
$$a = \sigma(z)$$

where z is the weighted sum, ω are the weights, x_i are the inputs, b is the bias, and σ is the activation function.

Training Process The training process involves forward propagation to compute outputs, loss calculation to measure the error, and backpropagation to update weights based on the error gradients. This process iterates over many epochs until the model converges.

Application Neural networks have been employed to develop real-time predictors for the spread of COVID-19, using geo-location and historical data to estimate daily case numbers with high accuracy, supporting effective pandemic management (Wieczorek et al. 2020).

5.1.18 Autoencoders

Origin Autoencoders were developed in the 1980s as unsupervised neural networks for learning compressed representations of data (Bourlard and Kamp 1988; Baldi and Hornik 1989). Key advancements were later introduced by Geoffrey Hinton and Richard Zemel in 1993, particularly in improving training techniques and their applications (Hinton and Zemel 1993).

Purpose Autoencoders are neural networks used for unsupervised learning to efficiently encode data into a lower-dimensional representation. The original input is then reconstructed from this representation. They are primarily used for dimensionality reduction and feature learning.

Mechanism An autoencoder consists of an encoder, which compresses the input into a lower-dimensional code, and a decoder, which reconstructs the input. The network is trained to minimize the reconstruction error between the input and the output.

Mathematical Formula
Let x be the input, z be the encoded representation, and $\hat{x}$ be the reconstructed input.

$$z = f(x)$$
$$\hat{x} = g(z)$$

where f and g are the encoding and decoding functions, respectively. The objective is to minimize the reconstruction loss:

$$L(x,\hat{x}) = \|x - \hat{x}\|^2$$

Training Process The autoencoder is trained using backpropagation to minimize the reconstruction error. The encoder learns to capture the essential features of the input, while the decoder learns to reconstruct the input from these features. This process iterates until the reconstruction error is minimal.

Application Autoencoders, specifically a variational autoencoder model, have been applied to COVID-19 detection and classification, integrating adaptive filtering and feature extraction techniques to achieve exceptional diagnostic accuracy (Mansour et al. 2021).

5.1.19 Convolutional Neural Networks (CNNs)

Origin CNNs were introduced by Yann LeCun and colleagues in 1989 to recognize handwritten postal codes, marking a milestone in computer vision (LeCun et al. 1989).

Purpose CNNs are designed to process and analyze visual data. They are particularly effective for tasks like image classification, object detection, and image segmentation.

Mechanism CNNs use convolutional layers to apply filters to input images, extracting hierarchical features. These layers are followed by pooling layers, which reduce dimensionality, and fully connected layers for final classification.

Mathematical Formula
For a convolutional layer:

$$(X * W)[i,j] = \Sigma_m \Sigma_n X[i+m, j+n] \cdot W[m,n]$$

where X is the input image, W is the filter, and $*$ denotes the convolution operation.

Training Process The training process involves forward propagation to compute feature maps, loss calculation to measure error, and backpropagation to update filter weights using error gradients. This process iterates over many epochs until the model converges.

Application CNNs have been utilized in a transfer learning framework for pneumonia detection in chest X-ray images, incorporating data augmentation techniques to enhance both training and validation performance, achieving superior diagnostic accuracy (Nath and Choudhury 2020).

5.1.20 Recurrent Neural Networks (RNNs)

Origin RNNs were conceptualized in the early 1980s, with John Hopfield introducing Hopfield networks, a recurrent architecture for associative memory (Hopfield 1982). The modern concept of RNNs for modeling dynamic temporal behaviors in sequential data was further developed by David Rumelhart and his colleagues in 1986 (Rumelhart et al. 1986).

Purpose RNNs are designed to handle sequential data and learn temporal dependencies. They are used for tasks like time series prediction, speech recognition, and language modeling.

Mechanism RNNs maintain a hidden state that captures information from previous time steps. They use this hidden state, along with the current input, to make predictions for the next time step.

Mathematical Formula
For a basic RNN cell:

$$h_t = \sigma\left(W_h h_{t-1} + W_x x_t + b\right)$$
$$y_t = \phi\left(V h_t\right)$$

where h_t is the hidden state, x_t is the input at time step t, W_h and W_x are weight matrices, b is the bias, σ is the activation function, and ϕ is the output function.

Training Process The training process involves forward propagation through time to compute outputs, loss calculation to measure error, and backpropagation through time (BPTT) to update weights based on error gradients. This process iterates over many epochs until the model converges.

Applications RNN frameworks, integrating LSTM architectures, have been developed for early detection and classification of lung diseases, including pneumonia and COVID-19, using advanced pre-processing and adaptive region of interest estimation techniques (Goyal and Singh 2023).

5.1.21 Long Short-Term Memory (LSTM) Networks

Origin LSTM networks were introduced by Sepp Hochreiter and Jürgen Schmidhuber in 1997 to address the vanishing gradient problem in RNNs (Hochreiter and Schmidhuber 1997).

Purpose LSTMs are a type of RNN designed to capture long-term dependencies in sequential data. They are particularly effective for tasks with long-range temporal dependencies, such as language translation and speech synthesis.

Mechanism LSTMs use memory cells and gating mechanisms to regulate the flow of information. The gates (input, forget, and output gates) control what information is kept, forgotten, or output at each time step.

Mathematical Formula
For an LSTM cell:

$$f_t = \sigma\left(W_f \cdot \left[h_{t-1,} x_t\right] + b_f\right)$$
$$i_t = \sigma\left(W_i \cdot \left[h_{t-1,} x_t\right] + b_i\right)$$
$$o_t = \sigma\left(W_o \cdot \left[h_{t-1,} x_t\right] + b_o\right)$$
$$\tilde{C}_t = \tan h\left(W_c \cdot \left[h_{t-1,} x_t\right] + b_c\right)$$

$$C_t = f_t \odot C_{t-1} + i_t \odot \tilde{C}_t$$
$$h_t = o_t \odot \tan h\left(C_t\right)$$

where f_t, i_t, and o_t are the forget, input, and output gates, respectively; C_t is the candidate cell state, C_t is the cell state, h_t is the hidden state, and $\odot$ denotes element-wise multiplication.

Training Process The training process involves forward propagation through the LSTM cells to compute outputs, loss calculation to measure the error, and BPTT to update weights based on error gradients. This process iterates over many epochs until the model converges.

Application LSTM models have been applied to predict human-virus protein–protein interactions, leveraging sequence data and word2vec embeddings to achieve high classification accuracy, contributing to therapeutic development (Tsukiyama et al. 2021).

5.1.22 Generative Adversarial Networks (GANs)

Origin Generative Adversarial Networks (GANs) were introduced by Ian Goodfellow and colleagues in 2014, leveraging adversarial training to generate realistic synthetic data (Goodfellow et al. 2014).

Purpose GANs are designed to generate realistic data samples. They are used for image generation, data augmentation, and other creative tasks.

Mechanism GANs consist of two neural networks: a generator and a discriminator. The generator creates fake data samples, while the discriminator evaluates their authenticity. The two networks are trained adversarially: the generator aims to produce convincing data and the discriminator aims to distinguish between real and fake data.

Mathematical Formula
The objective function for G

$$\max_D V\left(D,G\right) = \mathbb{E}_{x\sim P_{\text{data}}(x)}\left[\log D\left(x\right)\right] + \mathbb{E}_{z\sim P_z(z)}\left[\log\left(1 - D\left(G\left(Z\right)\right)\right)\right]$$

where $D(x)$ is the discriminator's estimate of the probability that x is real, $G(Z)$ is the generator's output from random noise z, and $P_{\text{data}}(x)$ and $P_z(z)$ are the distributions of real data and noise, respectively.

Training Process The training process involves alternately updating the generator and discriminator. The generator is trained to minimize the probability of the discriminator correctly identifying fake samples, while the discriminator is trained to maximize this probability. This adversarial training continues until a balance is achieved.

Application GANs have been applied to generate novel antiviral peptides, leveraging curated datasets to accelerate therapeutic development through the efficient design of peptide drug backbones (Surana et al. 2023).

5.1.23 Transformers

Origin Transformers were introduced by Ashish Vaswani and colleagues in 2017. Their self-attention mechanism revolutionized natural language processing by enabling parallel data processing (Vaswani et al. 2017).

Purpose Transformers are designed for processing sequential data, particularly in natural language processing tasks. They excel at tasks such as machine translation, text summarization, and language modeling.

Mechanism Transformers use self-attention mechanisms to weigh the importance of different words in a sequence, allowing for the parallel processing of input data. They consist of an encoder to process the input sequence and a decoder to generate the output sequence.

Mathematical Formula
For self-attention:

$$\text{Attention}(Q,K,V) = \text{softmax}\left(\frac{QK^T}{\sqrt{d_k}}\right)V$$

where Q (queries), K (keys), and V (values) are the input matrices, and d_k is the dimension of the keys.

Training Process The training process involves feeding input sequences to the encoder and training the decoder to predict the next word in the sequence. The model is optimized using backpropagation to minimize the difference between predicted and actual sequences. This process iterates over many epochs until the model converges.

Application Transformers, specifically a Vision Transformer-based deep learning pipeline, have been employed for COVID-19 detection in chest X-ray images, providing interpretable visualizations and demonstrating improved diagnostic accuracy compared to existing models (Shome et al. 2021).

5.2 Applications of AI in Virology Research

AI has emerged as a transformative tool in virology, offering sophisticated methods for analyzing large-scale datasets, predicting viral dynamics, and accelerating the development of antiviral therapeutics and vaccines. By leveraging a range of AI algorithms, researchers have significantly enhanced virus classification, drug discovery, diagnostics, and epidemiological modeling. This section explores key applications of AI in virology, illustrating its potential to revolutionize the field.

5.2.1 Virus Classification

Classification is essential for understanding the genetic diversity, evolutionary pathways, and epidemiological dynamics of different viruses. AI algorithms can substantially improve the precision and efficiency of classification by analyzing vast genomic datasets. Models such as SVMs, decision trees, and neural networks excel in identifying genetic markers and structural features that differentiate viral strains. Learning from labeled datasets, where each sample is associated with a known virus type, these models can classify novel or unknown viruses with remarkable accuracy. Ensemble methods, such as RFs and boosting, further improve robustness by combining the predictions of multiple models. AI tools can also reconstruct phylogenetic relationships among viruses, providing insights into their evolutionary history and tracing ancestral strains. Furthermore, real-time mutation tracking facilitated by AI contributes to keeping vaccines up-to-date and refining antiviral strategies, ensuring effective responses to rapidly evolving viral threats. These advances underscore the importance of AI in the early detection and surveillance of emerging pathogens.

5.2.2 Antiviral and Vaccine Development

AI has redefined the landscape of antiviral and vaccine development, offering innovative approaches to address viral threats with unprecedented speed and accuracy. Machine learning models, including deep learning and reinforcement learning, can simulate molecular interactions of drug candidates and predict their binding affinities to viral proteins, streamlining the identification of promising compounds. For antiviral drug development, AI can efficiently screen extensive chemical libraries, identifying compounds with the potential to inhibit viral replication. Predictions based on molecular structures and protein interactions prioritize candidates for experimental validation, conserving resources and time. In vaccine development, AI assists in identifying immunogenic epitopes—key regions of viral proteins that elicit immune responses—facilitating the design of peptide-based vaccines. AI also optimizes vaccine formulations by predicting the stability and efficacy of vaccine candidates under varying conditions; this significantly reduces development timelines while enhancing therapeutic potential.

5.2.3 Diagnostic Advancements

AI has transformed diagnostics in virology, enabling rapid and precise detection of infections. Neural network-based algorithms excel at analyzing genetic sequences, medical images, and clinical data, identifying infections with high sensitivity and specificity. For instance, CNNs can process complex biological datasets to identify patterns indicative of viral presence. AI-powered diagnostic systems can analyze samples such as blood or saliva to detect viral RNA or proteins, expediting diagnostic workflows and enabling timely patient care. By integrating diverse data sources—including imaging scans, laboratory results, and patient history—AI enhances diagnostic reliability. In pandemic scenarios, AI-driven diagnostics support large-scale screening and surveillance, facilitating the rapid containment of outbreaks. Predictive models further support public health efforts by forecasting infection dynamics and optimizing resource allocation.

5.2.4 Predictive Modelling and Epidemiology

AI has become indispensable in predictive modeling and epidemiology, providing tools to forecast viral transmission and evaluate the efficacy of public health interventions. Machine learning models analyze variables such as population density, mobility patterns, and environmental factors to predict transmission dynamics and assess outbreak scenarios. During the COVID-19 pandemic, AI-driven models offered critical insights into disease progression, informing decisions on lockdowns, vaccination campaigns, and resource distribution. These models simulate the potential outcomes of various intervention strategies, enabling policymakers to implement evidence-based measures to mitigate spread. Furthermore, AI can identify potential hotspots of future outbreaks by analyzing historical data and current trends. This proactive approach enables early intervention and targeted surveillance, reducing the risk of widespread transmission.

5.2.5 Personalized Medicine and Treatment Optimization

AI facilitates personalized medicine by tailoring treatments to individual patients based on their genetic profiles and medical histories. This precision approach can help address the variability in patient responses to antiviral therapies. Machine learning algorithms can be used to identify biomarkers that predict how likely a patient is to benefit from a specific treatment, enabling healthcare providers to select the most effective options. By monitoring patient responses and dynamically adjusting therapies, AI can also continuously optimize treatment regimens. For instance, AI-driven systems to track viral load and immune responses in real time can provide actionable insights for clinicians to modify treatment plans. By enhancing therapeutic efficacy and reducing adverse reactions, AI contributes to improved patient outcomes and more efficient use of healthcare resources.

5.2.6 Research and Discovery

AI accelerates research and discovery in virology by automating data analysis and uncovering novel insights, particularly from complex datasets. Machine learning models can process vast amounts of scientific literature, genomic data, and experimental results, identifying previously unrecognized viral genes, mutations, or pathways associated with virulence and drug resistance. These discoveries pave the way for new diagnostic markers, therapeutic targets, and vaccines. AI also facilitates collaboration by integrating data from diverse sources, enabling comprehensive models of viral behavior and host-pathogen interactions. This integrative approach fosters a deeper understanding of viral mechanisms and accelerates the pace of scientific discovery, driving innovation in diagnostics, therapeutics, and epidemiology.

The application of AI in virology research marks a paradigm shift in understanding, diagnosing, and combating viral diseases. By leveraging advanced algorithms, researchers have achieved breakthroughs in virus classification, antiviral and vaccine development, diagnostics, predictive modeling, and personalized medicine. As AI continues to evolve, its integration with virology promises even greater advances. These will lead to more effective strategies for combating viral infections and improving public health outcomes worldwide.

The following case study illustrates how machine learning can reveal key insights into the host-specific infection properties of polyomaviruses (PyVs) (Cho et al. 2021). In analyzing the large tumor antigen (LT-Ag) and capsid VP1 proteins—key markers for replication and host interaction—researchers used machine learning to classify PyVs based on host species, using kNN, RFs, and NB algorithms. This study demonstrates the power of machine learning to enhance our understanding of viral host specificity and lays the groundwork for predictive tools to study viral tropism and pathogen behavior.

5.3 Case Study: Machine Learning-Based Analysis of Protein Determinants Influencing Host-Specific Infection Characteristics of Polyomaviruses

PyVs are non-enveloped viruses with double-stranded DNA (dsDNA) genomes of approximately 5000 base pairs. The genome encodes at least two regulatory proteins—LT-Ag and small tumor antigen—and two structural capsid proteins—VP1 and VP2—with well-defined functions (DeCaprio and Garcea 2013; Moens et al. 2017). LT-Ag is highly conserved across PyV species, with specific functional domains that interact with the key tumor suppressor proteins, p53 and retinoblastoma (Rb). These interactions lead to disruptions in the cell cycle which can result in tumor formation (Borchert et al. 2014). Interestingly, in some avian and fish PyVs, LT-Ag proteins exhibit unique characteristics that suggest additional roles in these hosts. The capsid protein VP1 exhibits antigenic diversity within short, variable amino acid regions but the remaining sequences are conserved, making it a key marker for viral classification (Moens et al. 2017).

PyVs exhibit species-specific tendencies, much like papillomaviruses (Van Doorslaer

2013; Buck et al. 2016), often leading to varying levels of host toxicity. Avian PyVs are associated with acute and chronic diseases (Bernier et al. 1981; Bozeman et al. 1981; Johne and Müller 1998; Johne et al. 2006), particularly in young birds (Johne and Muller 2007). For instance, goose hemorrhagic PyV causes high mortality in young geese due to hemorrhagic nephritis and enteritis (Guerin et al. 2000). This marked pathogenicity contrasts with the relatively benign nature of mammalian PyVs, which typically cause subclinical infections in immunocompetent hosts (Krumbholz et al. 2009; Moens et al. 2017). However, mammalian PyVs exhibit oncogenic properties, whereas avian PyVs do not.

The widespread presence of PyVs and PyV-like sequences across a wide range of species—including vertebrates, invertebrates, and other organisms—underscores the need to investigate their host-specific determinants (Buck et al. 2016). To this end, LT-Ag and VP1 proteins provide critical insights into the mechanisms driving host specificity, virulence, and tumorigenicity (Moens et al. 2017). This case study aimed to classify PyVs by their host species using machine learning approaches, leveraging the physicochemical properties of LT-Ag and VP1 amino acid sequences. Furthermore, it sought to identify the unique characteristics of these proteins across host groups, thereby enhancing our understanding of PyV-host interactions.

To investigate host-specific properties of PyVs, LT-Ag and VP1 sequences from 86 species in the *Polyomaviridae* family were collected from the ICTV (https://talk.ictvonline.org/ictv-reports/ictv_online_report/dsdna-viruses/w/polyomaviridae) and NCBI GenBank databases (https://www.ncbi.nlm.nih.gov). These sequences represent a broad range of hosts, including mammals, birds, and fish. Multiple sequence alignments (MSAs) were conducted using the MUltiple Sequence Comparison by Log-Expectation (MUSCLE) software, which is widely recognized for its high accuracy and computational efficiency in aligning protein sequences through iterative refinement and advanced algorithmic techniques (Edgar 2004). Phylogenetic trees were constructed using the maximum likelihood method based on the Tamura-Nei model (Tamura and Nei 1993) in MEGA 7.0.26 software (Kumar et al. 2016). Bootstrapping with 1000 replicates was used to assess the reliability of the tree topologies (Felsenstein 1985). Phylogenetic analysis revealed distinct clustering patterns for LT-Ag and VP1. LT-Ag sequences grouped strongly according to host species—mammals (Group M), birds (Group A), and fish (Group F)—reflecting evolutionary and functional similarities within these groups (Fig. 5.1). Conversely, VP1 sequences exhibited more scattered clustering patterns, highlighting their role in receptor interaction and antigenic variability. These differences underscore the conserved nature of LT-Ag compared to the adaptability of VP1, reflecting their distinct evolutionary pressures.

Protein sequences of PyVs were transformed into numerical feature vectors to facilitate machine learning-based classification. This transformation was accomplished using the PROFEAT web server (Li et al. 2006). This tool computes physicochemical properties from amino acid sequences, including hydrophobicity, normalized van der Waals volume, polarity, charge, solvent accessibility, and other structural characteristics (Table 5.1).

These physicochemical attributes are critical for representing the biochemical properties and functional dynamics of viral proteins. To further enhance the feature representation, the composition, transition, and distribution (CTD) method was applied (Dubchak et al. 1995, 1997). This approach breaks down the protein sequence into three key descriptors: composition (C), which quantifies the proportion of amino acids with a particular property; transition (T), which measures the frequency of shifts between different properties; and distribution (D), which assesses the positional spread of amino acids with a specific property throughout the sequence. Together, these descriptors capture the fundamental characteristics of protein structure and function, making them highly informative for classification tasks. Each protein sequence was converted into 146 feature vectors, encompassing 21 attributes for each of six major physicochemical properties.

These features were subsequently used as inputs for machine learning analysis. This digitization process enabled systematic exploration of sequence-function relationships and provided a robust basis for predicting the host-specific infection characteristics of PyVs.

To evaluate the host classification potential of LT-Ag and VP1 protein sequences, three machine learning algorithms—kNN, RF, and NB—were implemented using WEKA 3.8.0 software (Witten et al. 1999). A ten-fold cross-validation approach was employed to divide the dataset into training and testing subsets, ensuring that the performance evaluation was unbiased and robust. These machine learning algorithms were chosen based on their ability to handle multi-class classification problems and their proven effectiveness in biological sequence analysis. Feature selection was a critical step in optimizing the performance of the classification models. Using correlation attribute evaluation, the physicochemical properties most strongly correlated with host classification were ranked by their Pearson correlation coefficients (Hall and Holmes 2003; Hall 1999, 2000). This process identified attributes with the most significant contributions to classification accuracy, while less relevant or redundant features were excluded. Feature subsets of varying sizes (5, 10, 25, 50, 75, 100, and 125 attributes) were constructed to systematically evaluate the impact of feature selection on model performance.

Among the three machine learning algorithms evaluated, kNN demonstrated the highest classification accuracy for both proteins, with 98.83% accuracy for LT-Ag. Additional impressive performance metrics included precision (0.989), recall (0.988), Matthews correlation coefficient (MCC = 0.951), and area under the curve (AUC = 0.974). For VP1, kNN also outperformed RF and NB approaches, achieving an accuracy of 96.51%, precision of 0.965, recall of 0.965, MCC of 0.852, and AUC of 0.944 (Table 5.2). These results highlight the robustness of kNN in capturing the evolutionary and functional distinctions between host groups based on viral protein sequences.

Optimal feature subsets were identified to improve computational efficiency and model performance. For LT-Ag, the inclusion of 25 features was sufficient to achieve the highest classification accuracy, reflecting the protein's conserved nature and strong correlation with host-specific properties. In contrast, VP1 required 75 included features to reach optimal performance, underscoring its greater sequence variability and functional complexity (Table 5.3 and Fig. 5.2). These findings emphasize the distinct roles of LT-Ag and VP1 in host classification, with LT-Ag serving as a more reliable marker due to its conserved functional domains; meanwhile, VP1 provides additional insights into receptor interaction and host specificity. LT-Ag was consistently associated with higher classification accuracies across all feature subsets, compared to VP1. However, classification using LT-Ag with fewer than 25 features failed to accurately classify fish PyVs, indicating the need for sufficient feature representation to capture interspecies variability.

Fig. 5.1 Reconstructed phylogenetic trees of 86 polyomaviruses (PyVs). PyVs were clustered according to their host species (mammal, avian, and fish) in phylogenetic trees constructed based on the amino acid sequences of the large tumor antigen (LT-Ag) (left) and VP1 (right) proteins. Symbols represent viral genera: [○] Alphapolyomaviruses; [▼] Betapolyomaviruses; [△] Deltapolyomaviruses; [■] Gammapolyomaviruses; [◆] Unassigned

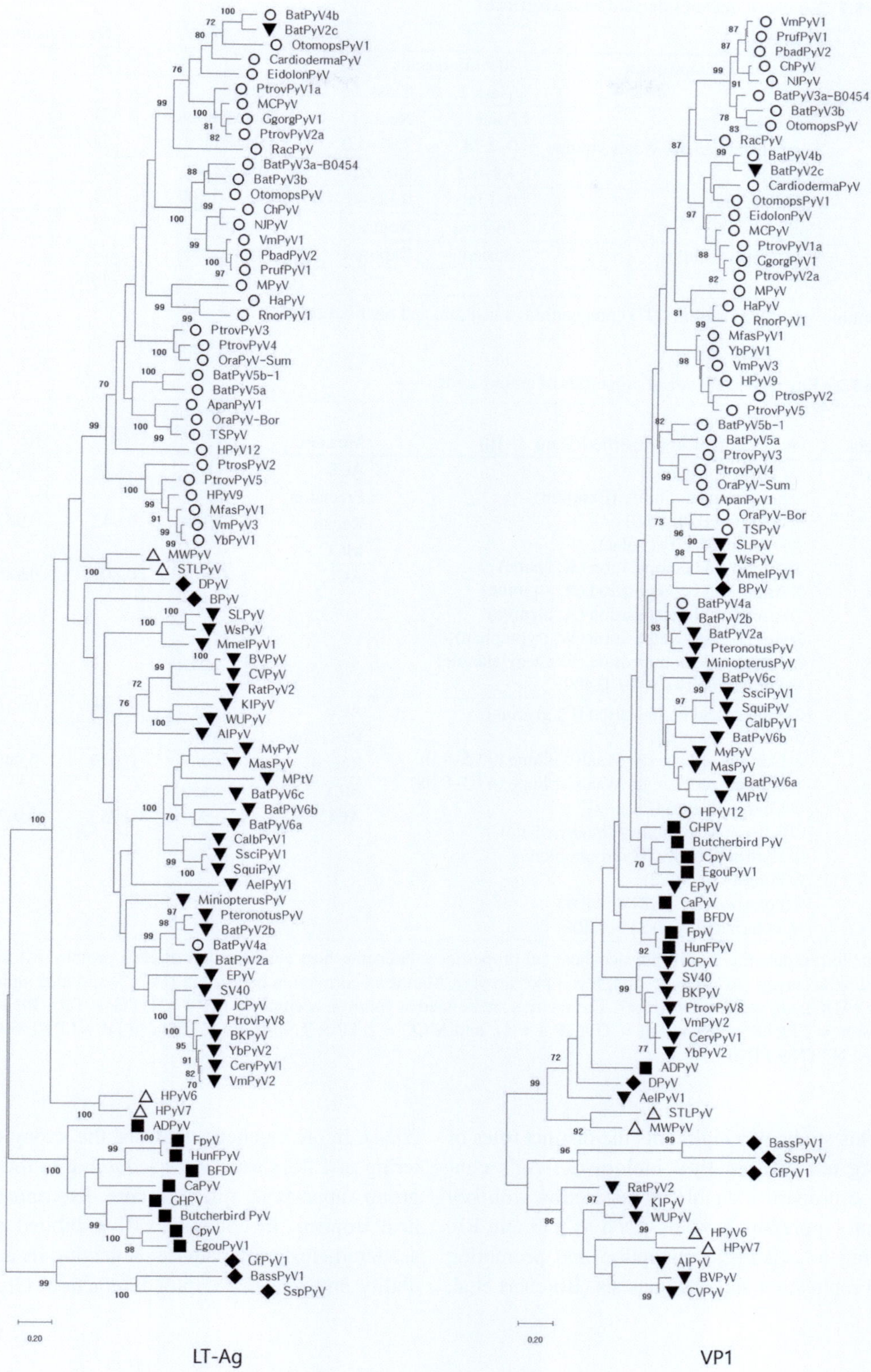
BatPyV4b
BatPyV2c
OtomopsPyV1
CardiodermaPyV
EidolonPyV
PtrovPyV1a
MCPyV
GgorgPyV1
PtrovPyV2a
RacPyV
BatPyV3a-B0454
BatPyV3b
OtomopsPyV
ChPyV
NJPyV
VmPyV1
PbadPyV2
PrufPyV1
MPyV
HaPyV
RnorPyV1
PtrovPyV3
PtrovPyV4
OraPyV-Sum
BatPyV5b-1
BatPyV5a
ApanPyV1
OraPyV-Bor
TSPyV
HPyV12
PtrosPyV2
PtrovPyV5
HPyV9
MfasPyV1
VmPyV3
YbPyV1
MWPyV
STLPyV
DPyV
BPyV
SLPyV
WsPyV
MmelPyV1
BVPyV
CVPyV
RatPyV2
KIPyV
WUPyV
AIPyV
MyPyV
MasPyV
MPtV
BatPyV6c
BatPyV6b
BatPyV6a
CalbPyV1
SsciPyV1
SquiPyV
AelPyV1
MiniopterusPyV
PteronotusPyV
BatPyV2b
BatPyV4a
BatPyV2a
EPyV
SV40
JCPyV
PtrovPyV8
BKPyV
YbPyV2
CeryPyV1
VmPyV2
HPyV6
HPyV7
ADPyV
FpyV
HunFPyV
BFDV
CaPyV
GHPV
Butcherbird PyV
CpyV
EgouPyV1
GfPyV1
BassPyV1
SspPyV
0.20
LT-Ag
VmPyV1
PrufPyV1
PbadPyV2
ChPyV
NJPyV
BatPyV3a-B0454
BatPyV3b
OtomopsPyV
RacPyV
BatPyV4b
BatPyV2c
CardiodermaPyV
OtomopsPyV1
EidolonPyV
MCPyV
PtrovPyV1a
GgorgPyV1
PtrovPyV2a
MPyV
HaPyV
RnorPyV1
MfasPyV1
YbPyV1
VmPyV3
HPyV9
PtrosPyV2
PtrovPyV5
BatPyV5b-1
BatPyV5a
PtrovPyV3
PtrovPyV4
OraPyV-Sum
ApanPyV1
OraPyV-Bor
TSPyV
SLPyV
WsPyV
MmelPyV1
BPyV
BatPyV4a
BatPyV2b
BatPyV2a
PteronotusPyV
MiniopterusPyV
BatPyV6c
SsciPyV1
SquiPyV
CalbPyV1
BatPyV6b
MyPyV
MasPyV
BatPyV6a
MPtV
HPyV12
GHPV
Butcherbird PyV
CpyV
EgouPyV1
EPyV
CaPyV
BFDV
FpyV
HunFPyV
JCPyV
SV40
BKPyV
PtrovPyV8
VmPyV2
CeryPyV1
YbPyV2
ADPyV
DPyV
AelPyV1
STLPyV
MWPyV
BassPyV1
SspPyV
GfPyV1
RatPyV2
KIPyV
WUPyV
HPyV6
HPyV7
AIPyV
BVPyV
CVPyV
0.20
VP1

Table 5.1 Set of 146 features derived from sequences

Feature	Property				No. of features
AAC	Amino acid composition	20 Amino acids			20
CTD		Class 2	Class 2	Class 3	
	Hydrophobicity	Polar	Neutral	Hydrophobicity	21
	Normalized van der Waals volume	0–2.78	2.95–4.0	4.03–8.08	21
	Polarity	4.9–6.2	8.0–9.2	10.4–13.0	21
	Polarizability	0–1.08	0.128–0.186	0.219–0.409	21
	Charge	Positive	Neutral	Negative	21
	Solvent accessibility	Buried	Exposed	Intermediate	21
Total					146

AAC amino acid composition, *CTD* composition, transition, and distribution descriptor

Table 5.2 Key physicochemical properties of amino acids

			Classifier		
Protein	Physicochemical properties (Rank 1~10)	Measure	kNN	NB	RF
LT-Ag	① Charge (Negative) ② Solvent accessibility (Exposed) ③ Polarity (10.4–13.0) ④ Hydrophobicity (Polar) ⑤ Amino acid composition (K; lysine) ⑥ Amino acid composition (R; arginine) ⑦ Amino acid composition (A; alanine) ⑧ Amino acid composition (W; tryptophan) ⑨ Amino acid composition (F; phenylalanine) ⑩ Polarizability (0.219–0.409)	ACC	98.83	94.18	95.34
		Precision	0.989	–	–
		Recall	0.988	0.942	0.953
		MCC	0.951	–	–
		AUC	0.974	0.801	0.988
VP1	① Amino acid composition (G; glycine) ② Charge (Neutral) ③ Normalized van der Waals volume (2.95–4.0) ④ Normalized van der Waals volume (4.03–8.08) ⑤ Charge (Neutral) ⑥ Solvent accessibility (Intermediate) ⑦ Hydrophobicity (Hydrophobicity) ⑧ Polarity (4.9–6.2) ⑨ Polarizability (0.219–0.409) ⑩ Polarizability (0.219–0.409)	ACC	96.51	91.86	94.18
		Precision	0.965	–	–
		Recall	0.965	0.919	0.942
		MCC	0.852	–	–
		AUC	0.944	0.752	0.977

The table presents the top 10 physicochemical properties influencing host classification of each protein. All values, including accuracy (ACC), precision, recall (sensitivity), Matthews correlation coefficient (MCC), and area under the curve (AUC), are weighted averages. The metrics are defined as follows: accuracy = (TP + TN)/(TP + TN + FP + FN); precision = TP/(TP + FP); recall = TP/(TP + FN); and MCC = {(TP × TN) − (FP × FN)}/√((TP + FP)(TN + FN)(TP + FN)(TN + FP))

This study also highlights the distinct roles of LT-Ag and VP1 in PyV biology. LT-Ag's conserved properties enable its interaction with key tumor suppressor proteins, such as p53 and Rb, leading to cell cycle disruption and promoting viral replication and oncogenesis (Borchert et al. 2014). In phylogenetic analysis, the strong clustering of LT-Ag sequences according to the host group supports its role as a core determinant of host tropism. In contrast, VP1 exhibited more scattered clustering patterns, reflecting its adaptability and role in receptor interaction. Glycine

Table 5.3 K-nearest neighbors model performance across various feature subsets

Protein	Measure	Feature subset 5	10	25	50	75	100	125	All features (146)
LT-Ag	ACC	91.86	96.51	98.83	97.67	98.83	98.83	98.83	98.83
	Precision	–	–	0.989	0.977	0.989	0.989	0.989	0.989
	Recall	0.919	0.965	0.988	0.977	0.988	0.988	0.988	0.988
	MCC	–	–	0.951	0.900	0.951	0.951	0.951	0.951
	AUC	0.801	0.922	0.974	0.948	0.974	0.974	0.974	0.974
VP1	ACC	91.86	93.02	93.02	96.51	97.67	97.67	95.34	96.51
	Precision	0.924	0.931	0.929	0.966	0.977	0.977	0.953	0.965
	Recall	0.919	0.930	0.930	0.965	0.977	0.977	0.953	0.965
	MCC	0.727	0.771	0.727	0.850	0.901	0.901	0.798	0.852
	AUC	0.901	0.867	0.840	0.922	0.948	0.948	0.918	0.944

Feature subsets were configured through ranking-based attribute selection for each protein. Accuracy (ACC), precision, recall, Matthews correlation coefficient (MCC), and area under the curve (AUC) values are presented as weighted averages

composition, charge, and van der Waals volume were identified as the most influential properties for VP1 classification. Amino acid substitutions in VP1 can alter receptor binding capacity, thereby influencing viral tropism and host range. For example, a single amino acid substitution at position 92 of VP1—glycine versus glutamic acid—has been shown to affect receptor binding, plaque size, and hemagglutination behavior in PyVs (Freund et al. 1991; Dubensky et al. 1991). These findings illustrate how VP1 adaptations contribute to host-specific infection characteristics and pathogenicity, despite being less strongly associated with host tropism.

Given the high classification accuracy achieved in this study, LT-Ag and VP1 proteins can be considered key parameters for classifying viral sequences, offering insight into host tropism and host-specific toxic effects of PyVs, respectively. These findings are likely influenced by biochemical functional variations across host species. To address variability in parameters, classifier accuracy was enhanced by reducing the impact of irrelevant attributes through feature selection. The kNN algorithm, optimized with feature subsets derived from ranking-based attribute selection, demonstrated a marked improvement in performance. Alterations in the amino acid sequences within the functional domains of viral proteins can significantly influence viral tropism, fitness, and host susceptibility. Further advances in this approach can be achieved by expanding the dataset to include a broader range of protein sequences and host types. Such developments could transform this method into a practical tool for directly predicting viral infection properties from their protein sequences. Accurate classification and prediction data on PyV infections are essential for improving disease prevention and treatment strategies. Acquiring detailed information about the pathogenic or infection-related properties of various PyV species would bolster clinical research aimed at developing preventive and therapeutic measures using classification and prediction models. Enhancing the precision of these models will require the identification of fine differences in amino acid attributes, including functional constraints at specific sequence positions. Addressing these intricacies is critical for deepening our understanding of viral protein sequence data and its application in predicting host interactions and disease outcomes.

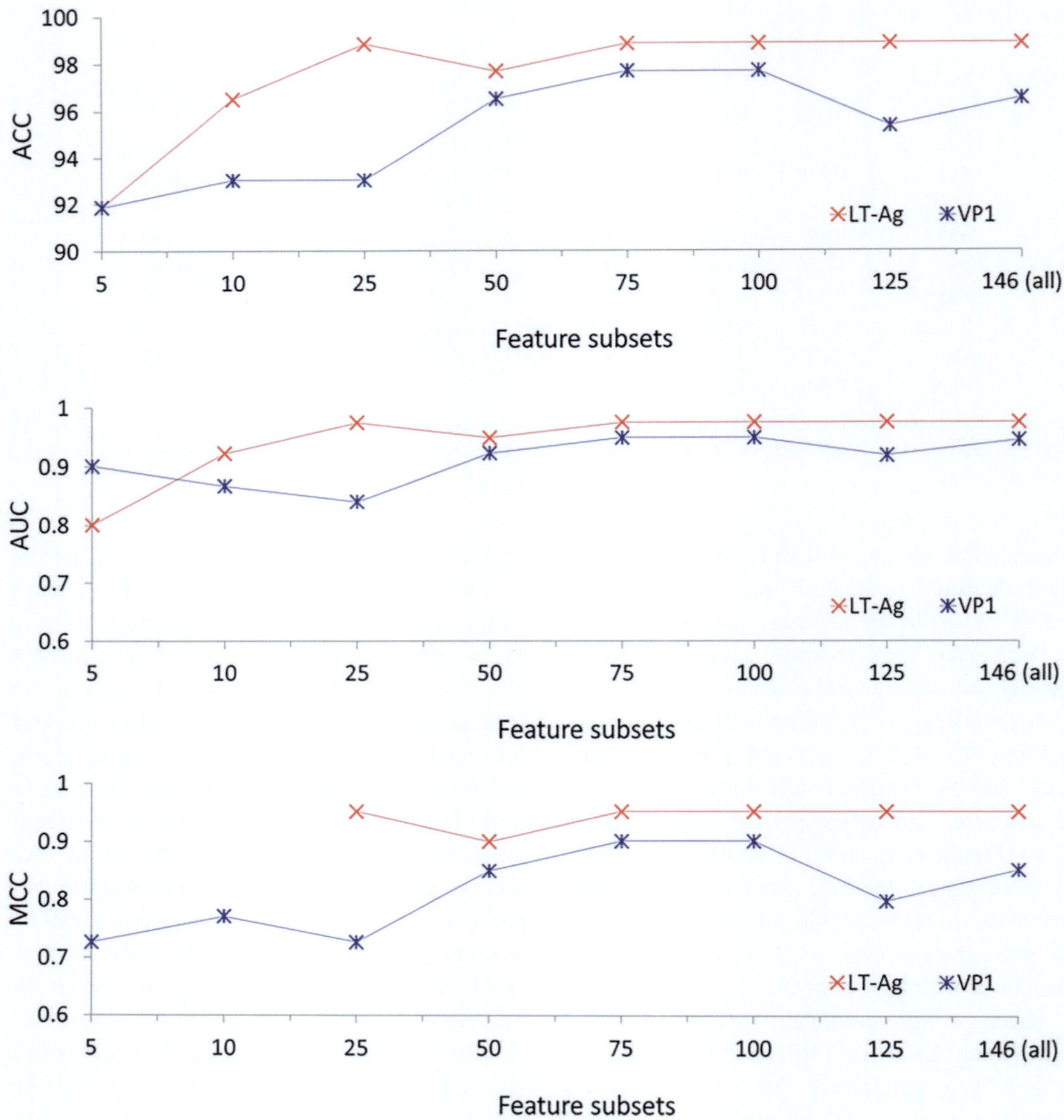

Fig. 5.2 Performance analysis of k-nearest neighbors (kNN) classifiers across feature subsets. The kNN algorithm—the top performer among the tested models—was applied to datasets with feature subsets ranked by attribute importance. Large tumor antigen (LT-Ag) consistently achieved higher classification accuracy across subsets (5–146 features), except for subsets with 5 and 10 features, where fish PyVs could not be classified, preventing MCC calculation

5.4 Conclusion

AI has emerged as a transformative tool for addressing complex challenges across diverse domains. This chapter has explored the foundational principles of AI algorithms, their implementation methods, and their versatile applications, particularly in the field of virology. By examining learning paradigms, structural architectures, and the adaptability of AI algorithms to varied tasks and data types, we have highlighted how these tools can drive breakthroughs in research and practical problem-solving. AI's impact on virology research is particularly noteworthy. From advancing virus classification, accelerating antiviral and vaccine development, and enhancing diagnostic precision and predictive epidemiology, AI has significantly advanced our ability to combat viral threats. Through the precise analysis of viral genome sequences, prediction of host-virus interactions, and optimization of public health strategies, machine learning models have become indispensable in tackling the complexities of infectious diseases. These contributions are especially critical in today's landscape, where rapid and effective responses to emerging infectious diseases are paramount.

The case study on PyVs showcases the practical application of AI in virology. By employing machine learning techniques to analyze protein determinants such as LT-Ag and VP1, the study revealed critical physicochemical properties that influence host specificity and pathogenicity. The superior performance of the kNN algorithm underscores the importance of rigorous feature selection and optimization in enhancing predictive accuracy. These insights deepen our understanding of viral behavior and host-virus dynamics, offering valuable contributions to disease prevention and therapeutic advancements.

Beyond this case study, AI's transformative potential in virology extends to a multitude of applications. AI-driven virus classification has revolutionized the analysis of expansive genomic datasets, enabling the rapid identification of viral strains and insights into their evolutionary relationships. Diagnostic advances powered by AI facilitate timely and accurate infection detection, while predictive models inform public health strategies to contain and mitigate outbreaks effectively. Furthermore, AI accelerates the discovery of antiviral drugs and vaccines by simulating molecular interactions, identifying therapeutic targets, and optimizing vaccine formulations. These advances not only enhance pandemic preparedness but also strengthen global health security over the long term.

Looking ahead, the future of AI in virology and other scientific domains hinges on expanding datasets, integrating more sophisticated algorithms, and incorporating detailed biological structures and dynamic interactions into predictive models. AI's ability to process and synthesize vast amounts of biological, environmental, and epidemiological data will further refine our understanding of complex systems, unlocking new opportunities for therapeutic interventions and epidemiological insights. The development of explainable AI (XAI) is particularly promising, as it will enhance the interpretability and transparency of machine learning models, enabling researchers and clinicians to better understand and trust the underlying mechanisms driving AI predictions.

This chapter reaffirms that AI is not just a computational tool but a transformative framework that can redefine how we approach, understand, and address complex challenges in the modern era. By bridging theoretical foundations with real-world applications, AI continues to revolutionize research, innovation, and problem-solving across a multitude of fields. Its integration with virology and other disciplines is poised to lead to groundbreaking discoveries, improve health outcomes, and guide effective strategies for addressing critical scientific and societal challenges. As AI evolves, its potential to shape the future of research, industry, and public health will remain unparalleled, marking it as an essential driver of progress in the twenty-first century.

References

Abdullah D, Susilo S, Ahmar AS, Rusli R, Hidayat R. The application of K-means clustering for province clustering in Indonesia of the risk of the COVID-19 pandemic based on COVID-19 data. Qual Quant. 2022;56:1283–91.

Afonso AM, Ebell MH, Gonzales R, Stein J, Genton B, Senn N. The use of classification and regression trees to predict the likelihood of seasonal influenza. Fam Pract. 2012;29:671–7.

Arslan H, Arslan H. A new COVID-19 detection method from human genome sequences using CpG island features and KNN classifier. Eng Sci Technol Int J. 2021;24:839–47.

Baldi P, Hornik K. Neural networks and principal component analysis: learning from examples without local minima. Neural Netw. 1989;2:53–8.

Bayes T. An essay towards solving a problem in the doctrine of chances. Philos Trans R Soc Lond. 1763;53:370–418.

Bernier G, Morin M, Marsolais G. A generalized inclusion body disease in the budgerigar (Melopsittacus undulatus) caused by a papovavirus-like agent. Avian Dis. 1981;25:1083–92.

Borchert S, Czech-Sioli M, Neumann F, Schmidt C, Wimmer P, Dobner T, Grundhoff A, Fischer N. High-affinity Rb binding, p53 inhibition, subcellular localization, and transformation by wild-type or tumor-derived shortened Merkel cell polyomavirus large T antigens. J Virol. 2014;88:3144–60.

Bourlard H, Kamp Y. Auto-association by multilayer perceptrons and singular value decomposition. Biol Cybern. 1988;59:291–4.

Bozeman LH, Davis RB, Gaudry D, Lukert PD, Fletcher OJ, Dykstra MJ. Characterization of a papovavirus isolated from fledgling budgerigars. Avian Dis. 1981;25:972–80.

Breiman L. Bagging predictors. Mach Learn. 1996;24:123–40.

Breiman L. Random forests. Mach Learn. 2001;45:5–32.

Buck CB, Van Doorslaer K, Peretti A, Geoghegan EM, Tisza MJ, An P, Katz JP, Pipas JM, McBride AA, Camus AC, McDermott AJ, Dill JA, Delwart E, Varsani A. The ancient evolutionary history of polyomaviruses. PLoS Pathog. 2016;12:e1005574.

Cho M, Kim H, Son HS. Analysis of protein determinants of host-specific infection properties of polyomaviruses using machine learning. Genes Genomics. 2021;43:407–20.

Cox DR. The regression analysis of binary sequences. J R Stat Soc B. 1958;20:215–32.

DeCaprio JA, Garcea RL. A cornucopia of human polyomaviruses. Nat Rev Microbiol. 2013;11:264–76.

Dubchak I, Muchnik I, Holbrook SR, Kim SH. Prediction of protein folding class using global description of amino acid sequence. Proc Natl Acad Sci USA. 1995;92:8700–4.

Dubchak I, Muchnik IB, Kim SH. Protein folding class predictor for SCOP: approach based on global descriptors. Proc Int Conf Intell Syst Mol Biol. 1997;5:104–7.

Dubensky TW, Freund R, Dawe CJ, Benjamin TL. Polyomavirus replication in mice: influences of VP1 type and route of inoculation. J Virol. 1991;65:342–9.

Edgar RC. MUSCLE: multiple sequence alignment with high accuracy and high throughput. Nucleic Acids Res. 2004;32:1792–7.

Felsenstein J. Confidence limits on phylogenies: an approach using the bootstrap. Evolution. 1985;39:783–91.

Feng PM, Ding H, Chen W, Lin H. Naive Bayes classifier with feature selection to identify phage virion proteins. Comput Math Methods Med. 2013;2013:530696.

Fix E, Hodges JL. Discriminatory analysis: nonparametric discrimination: consistency properties. Technical Report 4. Randolph Field, TX: USAF School of Aviation Medicine; 1951.

Freund Y, Schapire RE. A decision-theoretic generalization of on-line learning and an application to boosting. J Comput Syst Sci. 1997;55:119–39.

Freund R, Calderone A, Dawe CJ, Benjamin TL. Polyomavirus tumor induction in mice: effects of polymorphisms of VP1 and large T antigen. J Virol. 1991;65:335–41.

Gauss CF. Theoria motus corporum coelestium in sectionibus conicis solem ambientium. Besser: Friedrich Perthes und I.H; 1809.

Gomes ALV, Wee LJ, Khan AM, Gil LH, Marques ET Jr, Calzavara-Silva CE, Tan TW. Classification of dengue fever patients based on gene expression data using support vector machines. PLoS One. 2010;5:e11267.

Goodfellow I, Pouget-Abadie J, Mirza M, Xu B, Warde-Farley D, Ozair S, Courville A, Bengio Y. Generative adversarial nets. Adv Neural Inf Proces Syst. 2014;27.

Goyal S, Singh R. Detection and classification of lung diseases for pneumonia and COVID-19 using machine and deep learning techniques. J Ambient Intell Humaniz Comput. 2023;14:3239–59.

Guerin JL, Gelfi J, Dubois L, Vuillaume A, Boucraut-Baralon C, Pingret JL. A novel polyomavirus (goose hemorrhagic polyomavirus) is the agent of hemorrhagic nephritis enteritis of geese. J Virol. 2000;74:4523–9.

Hall MA. Correlation-based feature selection for machine learning. PhD thesis. New Zealand: University of Waikato, Department of Computer Science; 1999.

Hall MA. Correlation-based feature selection for discrete and numeric class machine learning. In: Proceedings of the 17th International conference on machine learning (ICML-2000); 2000. p. 359–66.

Hall MA, Holmes G. Benchmarking attribute selection techniques for discrete class data mining. IEEE Trans Knowl Data Eng. 2003;15:1437–47.

Hinton GE, Zemel R. Autoencoders, minimum description length and Helmholtz free energy. In: Proceedings of the 6th international conference on neural information processing systems; 1993. p. 3–10.

Hochreiter S, Schmidhuber J. Long short-term memory. Neural Comput. 1997;9:1735–80.

Hopfield JJ. Neural networks and physical systems with emergent collective computational abilities. Proc Natl Acad Sci USA. 1982;79:2554–8.

Hotelling H. Analysis of a complex of statistical variables into principal components. J Educ Psychol. 1933;24:417–41.

Jaimes JA, André NM, Chappie JS, Millet JK, Whittaker GR. Phylogenetic analysis and structural modeling of SARS-CoV-2 spike protein reveals an evolutionary distinct and proteolytically sensitive activation loop. J Mol Biol. 2020;432:3309–25.

Johne R, Müller H. Avian polyomavirus in wild birds: genome analysis of isolates from Falconiformes and Psittaciformes. Arch Virol. 1998;143:1501–12.

Johne R, Muller H. Polyomaviruses of birds: etiologic agents of inflammatory diseases in a tumor virus family. J Virol. 2007;81:11554–9.

Johne R, Wittig W, Fernández-de-Luco D, Höfle U, Müller H. Characterization of two novel polyomaviruses of birds by using multiply primed rolling-circle amplification of their genomes. J Virol. 2006;80:3523–31.

Johnson SC. Hierarchical clustering schemes. Psychometrika. 1967;32:241–54.

Krumbholz A, Bininda-Emonds OR, Wutzler P, Zell R. Phylogenetics, evolution, and medical importance of polyomaviruses. Infect Genet Evol. 2009;9:784–99.

Kumar S, Stecher G, Tamura K. MEGA7: molecular evolutionary genetics analysis version 7.0 for bigger datasets. Mol Biol Evol. 2016;33:1870–4.

Kumar R, Maheshwari S, Sharma A, Linda S, Kumar S, Chatterjee I. Ensemble learning-based early detection of influenza disease. Multimed Tools Appl. 2024;83:5723–43.

Lara-Ramírez EE, Rivera G, Oliva-Hernández AA, Bocanegra-Garcia V, López JA, Guo X. Unsupervised learning analysis on the proteomes of Zika virus. PeerJ Comput Sci. 2024;10:e2443.

LeCun Y, Boser B, Denker JS, Henderson D, Howard RE, Hubbard W, Jackel LD. Backpropagation applied to handwritten zip code recognition. Neural Comput. 1989;1:541–51.

Li ZR, Lin HH, Han LY, Jiang L, Chen X, Chen YZ. PROFEAT: a web server for computing structural and physicochemical features of proteins and peptides from amino acid sequence. Nucleic Acids Res. 2006;34:W32–7.

Li L, Chang D, Han L, Zhang X, Zaia J, Wan XF. Multitask learning sparse group lasso: a method for quantifying antigenicity of influenza A (H1N1) virus using mutations and variations in glycosylation of Hemagglutinin. BMC Bioinf. 2020;21:1–22.

Lloyd S. Least squares quantization in PCM. IEEE Trans Inf Theory. 1982;28:129–37.

Ma Y, Zhao Y, Liu J, He X, Wang B, Fu S, Yan J, Niu J, Ji Zhou J, Luo B. Effects of temperature variation and humidity on the death of COVID-19 in Wuhan, China. Sci Total Environ. 2020;724:138226.

Machado G, Mendoza MR, Corbellini LG. What variables are important in predicting bovine viral diarrhea virus? A random forest approach. Vet Res. 2015;46:1–15.

Mansour RF, Escorcia-Gutierrez J, Gamarra M, Gupta D, Castillo O, Kumar S. Unsupervised deep learning based variational autoencoder model for COVID-19 diagnosis and classification. Pattern Recogn Lett. 2021;151:267–74.

McCulloch WS, Pitts WA. A logical calculus of the ideas immanent in nervous activity. Bull Math Biophys. 1943;5:115–33.

Moens U, Krumbholz A, Ehlers B, Zell R, Johne R, Calvignac-Spencer S, Lauber C. Biology, evolution, and medical importance of polyomaviruses: an update. Infect Genet Evol. 2017;54:18–38.

Muhammad I, Mukhlash I, Jamhuri M, Iqbal M, Irawan MI. Classification of COVID-19 variants using boosting algorithm. In: 2022 9th international conference on Electrical Engineering, Computer Science and Informatics (EECSI), October. IEEE; 2022. p. 29–34.

Nath M, Choudhury C. Automatic detection of pneumonia from chest X-rays using deep learning. In: Machine learning, image processing, network security and data sciences: second international conference, MIND 2020, Silchar, India, July 30–31, 2020, proceedings, part I, vol. 2. Springer Singapore; 2020. p. 175–82.

Pearson K. On lines and planes of closest fit to systems of points in space. Philos Mag. 1901;2:559–72.

Quinlan JR. Induction of decision trees. Mach Learn. 1986;1:81–106.

Rosenblatt F. The perceptron: a probabilistic model for information storage and organization in the brain. Psychol Rev. 1958;65:386–408.

Rumelhart DE, Hinton GE, Williams RJ. Learning representations by back-propagating errors. Nature. 1986;323:533–6.

Shome D, Kar T, Mohanty SN, Tiwari P, Muhammad K, AlTameem A, Zhang Y, Saudagar AKJ. COVID-transformer: interpretable COVID-19 detection using vision transformer for healthcare. Int J Environ Res Public Health. 2021;18:11086.

Surana S, Arora P, Singh D, Sahasrabuddhe D, Valadi J. Pandoragan: generating antiviral peptides using generative adversarial network. SN Comput Sci. 2023;4:607.

Tamura K, Nei M. Estimation of the number of nucleotide substitutions in the control region of mitochondrial DNA in humans and chimpanzees. Mol Biol Evol. 1993;10:512–26.

Tibshirani R. Regression shrinkage and selection via the lasso. J R Stat Soc B. 1996;58:267–88.

Tsukiyama S, Hasan MM, Fujii S, Kurata H. LSTM-PHV: prediction of human-virus protein–protein interactions by LSTM with word2vec. Brief Bioinform. 2021;22:bbab228.

van der Maaten L, Hinton G. Visualizing data using t-SNE. J Mach Learn Res. 2008;9:2579–605.

Van Doorslaer K. Evolution of the papillomaviridae. Virology. 2013;445:11–20.

Vapnik VN, Chervonenkis AY. On a class of perceptrons. Autom Remote Control. 1964;25:103–9.

Vaswani A, Shazeer N, Parmar N, Uszkoreit J, Jones L, Gomez AN, Lukasz K, Polosukhin I. Attention is all you need. Adv Neural Inf Proces Syst. 2017;30:arXiv:1706.03762 [cs.CL].

Wang B, Jiang L. Principal component analysis applications in COVID-19 genome sequence studies. Cogn Comput. 2021;1-12:1637.

Wang P, Zhang W, Wang H, Shi C, Li Z, Wang D, Luo L, Du Z, Hao Y. Predicting the incidence of infectious diarrhea with symptom surveillance data using a stacking-based ensembled model. BMC Infect Dis. 2024;24:265.

Wieczorek M, Siłka J, Połap D, Woźniak M, Damaševičius R. Real-time neural network based predictor for cov19 virus spread. PLoS One. 2020;15:e0243189.

Witten IH, Frank E, Trigg LE, Hall MA, Holmes G, Cunningham SJ. Weka: practical machine learning tools and techniques with Java implementations

(Working paper 99/11). Hamilton: University of Waikato, Department of Computer Science; 1999.

Wolpert DH. Stacked generalization. Neural Netw. 1992;5:241–59.

Wynants L, Van Calster B, Collins GS, Riley RD, Heinze G, Schuit E, Bonten MMJ, Dahly DL, Damen JAA, Debray TPA, de Jong VMT, De Vos M, Dhiman P, Haller MC, Harhay MO, Henckaerts L, Heus P, Kammer M, Kreuzberger N, Lohmann A, Luijken K, Ma J, Martin GP, McLernon DJ, Andaur Navarro CL, Reitsma JB, Sergeant JC, Shi C, Skoetz N, Smits LJM, Snell KIE, Sperrin M, Spijker R, Steyerberg EW, Takada T, Tzoulaki I, van Kuijk SMJ, van Bussel B, van der Horst ICC, van Royen FS, Verbakel JY, Wallisch C, Wilkinson J, Wolff R, Hooft L, Moons KGM, van Smeden M. Prediction models for diagnosis and prognosis of COVID-19: systematic review and critical appraisal. BMJ. 2020;369:m1328.

Xie Y, Xu L, Li J, Deng X, Hong Y, Kolivras K, Gaines DN. Spatial variable selection and an application to Virginia Lyme disease emergence. J Am Stat Assoc. 2019;114:1–30.

Zou H, Hastie T. Regularization and variable selection via the elastic net. J R Stat Soc B. 2005;67:301–20.

Part III

Applications and Case Studies

6 Tracing the Driving Forces of Viral Evolution

Viruses are among the most rapidly evolving entities on Earth, constantly undergoing genetic changes that allow them to adapt to hosts and environmental pressures. The driving forces behind this evolution are multifaceted, shaped by high mutation rates, complex host interactions, immune evasion strategies, and transmission dynamics, among other factors. Unraveling these mechanisms is crucial not only for the understanding of viral biology but also for the design of effective antiviral therapies, vaccines, and public health strategies. The development and application of computational and statistical methods are vital for studying viral evolution, providing insights into genetic architecture, evolutionary trajectories, and selective pressures. Phylogenetic analysis, for instance, enables us to trace evolutionary relationships between viral strains, while compositional and codon usage analyses reveal patterns that shed light on evolutionary constraints and adaptations. Codon usage, in particular, plays a pivotal role in viral evolution, influencing the efficiency of protein synthesis, immune recognition, and cross-species transmission potential. These factors are crucial for understanding viral fitness, host specificity, and pathogenicity.

In this chapter, we will explore some key analytical methods used to study viral evolution, including those outlined in the *Essential Methods for Viroinformatics* section. These methods encompass phylogenetic analysis, nucleotide compositional analysis, and a comprehensive suite of codon usage analyses; these include relative synonymous codon usage (RSCU), the effective number of codons (ENC), parity rule 2 (PR2), neutrality, and selection pressure analyses, and the codon adaptation index (CAI) calculation. Each method offers a distinct perspective on viral adaptation and evolution, and collectively they contribute a deeper understanding of the genetic and environmental factors shaping viral genomes. This chapter presents two case studies to demonstrate the practical application of these methods. The first examines codon usage patterns in the oncogenes of human papillomavirus (HPV), a virus known for its role in cervical cancer (Cho et al. 2021). Through this analysis, we explore how codon usage biases in HPV oncogenes reflect viral adaptation to human host environments, potentially influencing carcinogenesis. The second case study investigates the codon usage patterns of large tumor antigen (LT-Ag) genes in PyVs across a range of host species (Cho et al. 2019). This analysis sheds light on host-specific adaptations, revealing the evolutionary strategies employed by PyVs to optimize replication and interaction within their respective hosts. Through these case studies, we aim to demonstrate the importance of codon usage as a critical evolutionary factor, while highlighting other key selective pressures and constraints

H. S. Son, M. Cho, *Viroinformatics and Computational Virology in Public Health*,
https://doi.org/10.1007/978-981-96-5757-5_6

shaping viral genomes. By leveraging computational approaches, these analyses provide valuable insights into the intricate dynamics of viral evolution, advancing our understanding of viral biology and informing future research. The exploration of these evolutionary dynamics has far-reaching implications, deepening our knowledge of viruses and facilitating improved control and prevention measures. By integrating these computational methods into virology research, we can better anticipate and respond to emerging viral threats, paving the way for innovative strategies in disease mitigation and public health.

6.1 Overview of Analytical Methods in Viral Evolution

6.1.1 Phylogenetic Analysis

Multiple sequence alignment is performed using MUSCLE (Edgar 2004), ClustalW (Chenna et al. 2003), and other tools to analyze the target sequences. Phylogeny is reconstructed using the maximum likelihood method, based on the Tamura-Nei model (Tamura and Nei 1993), or other methods such as Neighbor-Joining (Saitou and Nei 1987) and Bayesian Inference (Li 1996; Mau 1996; Mau and Newton 1997; Mau et al. 1999; Rannala and Yang 1996; Yang and Rannala 1997). These methods are implemented with software like MEGA (Kumar et al. 2016), PhyML (Guindon et al. 2010), and MrBayes (Huelsenbeck and Ronquist 2001). Bootstrap analysis is conducted with 1000 repetitions of the dataset to assess the robustness of individual nodes (Felsenstein 1985). The reconstructed tree can then be used to identify phylogenetic relationships for genes or protein sequences at the level of genes, strains, and species. Based on these results, various viral genes, strains, and species are grouped according to their phylogenetic relationships. This classification information is considered when interpreting and discussing the results of evolutionary studies, codon pattern analysis, and other related analyses.

6.1.2 Nucleotide Compositional Analysis

The analysis of nucleotide composition corresponding to the coding sequences (CDS) of genes or genomes can be performed using programs like CodonW (https://sourceforge.net/projects/codonw/) and CAIcal (http://genomes.urv.es/CAIcal/) (Puigbò et al. 2008). This analysis enables the calculation of nucleotide frequency (%A, %C, %T, and %G), GC and AT content (%GC and %AT), nucleotide frequency in the third position of synonymous codons (%A3, %C3, %T3, and %G3), G + C content in the third codon position (%GC3), A + T content in the third codon position (%AT3), and average G + C content (%GC12) or A + T content (%AT12) in the first and second codon positions. Additionally, genetic variability, the total number of segregating sites, the total number of mutations, the average number of nucleotide differences between sequences, and nucleotide diversity can be estimated using DnaSP (Librado and Rozas 2009).

6.1.3 Codon Usage Pattern Analyses

6.1.3.1 Relative Synonymous Codon Usage (RSCU) Analysis

RSCU is a measure of the preference for synonymous codon usage. It is defined as the ratio of observed synonymous codon usage to the expected value if each codon in a group of synonymous codons were used equally (Sharp and Li 1986a). Codons with an RSCU value greater than 1.0 are considered to have a positive codon usage bias, indicating a relatively higher preference (more abundant codon); codons with an RSCU value less than 1.0 are considered to have a negative codon usage bias, indicating a relatively lower preference (less abundant codon). An RSCU value of 1.0 suggests an equal preference for synonymous codons or that codon usage is random (Sharp and Li 1986b). Generally, codons with an RSCU value ≥ 1.6 are considered overrepresented, and those with an RSCU value ≤0.6 are considered underrepresented (Wong et al. 2010). The RSCU value of a gene CDS can be calculated using programs like CodonW and CAIcal, and the indices calculated for each sequence or set of sequences can be compared, analyzed, and visualized using tools like XLSTAT, R, and others.

6.1.3.2 Effective Number of Codons (ENC) Analysis

The ENC is calculated to quantify the absolute codon usage bias in the gene CDS, independent of gene length and without requiring a reference set. ENC value ranges from 20 to 61. A value of 20 indicates maximum codon usage bias, meaning only one of the possible synonymous codons is used for a particular amino acid. In contrast, a value of 61 indicates no bias, meaning all possible synonymous codons are used equally for the corresponding amino acid. Genes with an ENC value ≤35 are generally considered to have significant codon bias (Wright 1990; Comeron and Aguadé 1998).

6.1.3.3 Parity Rule 2 (PR2) Analysis

PR2 analysis is used to examine the effects of mutations and selection pressures on the codon usage bias of genes. The PR2 plot positions the third-codon position biases for AT [A3/(A3 + T3)] and GC [G3/(G3 + C3)] in fourfold degenerate codon families—Ala (A), Arg (R), Gly (G), Leu (L), Pro (P), Ser (S), Thr (T), and Val (V)—across the entire genome. The AT bias is represented on the vertical axis (y), and the GC bias on the horizontal axis (x). The position of the plot where both coordinates are 0.5 is A = T, G = C (PR2), indicating no bias between the effects of mutations and natural selection (substitution rate). The distance between the coordinate position (0.5, 0.5) and the plot point (the center of the plot) indicates the degree and direction of the PR2 bias (Sueoka 1995, 1999).

6.1.3.4 Neutrality Analysis

Neutrality analysis assesses the respective contributions of mutational pressure and natural selection in shaping codon usage bias within viral gene groups. Neutrality plots are used to explore the relationship between GC content at the first and second codon positions (GC12) and the third codon position (GC3). These plots provide insight into the balance between directional mutation pressure and natural selection, which govern evolutionary changes. The gradient of the regression line within a neutrality plot reflects the degree of neutrality, helping to determine the dominant evolutionary force. A gradient of 0 implies that directional mutation pressure has no impact, while a gradient of 1 indicates complete neutrality (Sueoka 1988). In the evaluation of genetic differentiation and gene flow, specific neutrality measures such as Tajima's D (Tajima 1989), Fu and Li's D*, and Fu and Li's F* (Fu and Li 1993) are calculated. These statistical tests offer insights into deviations from neutrality. Tajima's D compares the observed differences between the number of segregating sites and the average pairwise nucleotide differences, measuring whether mutations align with neutral evolution. Fu and Li's D* focuses on the disparity between the number of singletons (mutations occurring only once) and the total mutation count. Fu and Li's F* examines the difference between singletons and the average pairwise nucleotide differences across sequences, further illuminating evolutionary dynamics (Fu and Li 1993; Tsompana et al. 2005). For this neutrality analysis, sequence alignments are performed using tools such as Clustal Omega (Chenna et al. 2003) and MAFFT (Katoh et al. 2005). Subsequently, DnaSP (Librado and Rozas 2009) and MEGA (Kumar et al. 2016) can be employed to compute neutrality indices and genetic diversity metrics. These include the total number of segregating sites, the total mutation count, and nucleotide diversity. These indices enable a comprehensive understanding of the evolutionary pressures shaping the viral genome and help identify underlying genetic variability.

6.1.3.5 Selection Pressure Analysis

Selection pressure analysis quantifies the evolutionary forces acting on viral genes by evaluating the rates of synonymous (dS) and non-synonymous (dN) substitutions. This method offers insight into the type and intensity of selection that shapes viral evolution. The ratio of dN to dS (ω) serves as the primary measure of selection pressure, where $\omega > 1$ suggests positive (or diversifying) selection, $\omega = 1$ indicates neutral selection, and $\omega < 1$ reflects negative (or purifying) selection (Nei and Gojobori 1986; Kryazhimskiy and Plotkin 2008). Positive selection promotes genetic diversity by favoring advantageous mutations, while purifying selection eliminates deleterious mutations, maintaining genetic stability (Yang and Bielawski 2000). To perform selection pressure analysis, nucleotide sequences of viral genes are analyzed for all isolates within a given virus group. Tools like MEGA enable precise calculation of dN, dS, and the resulting dN/dS ratios (ω) for each viral gene or domain (Kumar et al. 2016). These ratios highlight the selective forces acting on specific genes and domains, allowing the identification of evolutionary patterns across subtypes and isolates. Selection pressure analysis provides essential information about the functional importance of viral genes by revealing whether mutations are driven by natural selection or random drift. Genes under positive selection suggest adaptation to environmental pressures or immune evasion. In contrast, genes under purifying selection reflect evolutionary constraints, indicating functional conservation (Kryazhimskiy and Plotkin 2008; Yang and Bielawski 2000). Neutral selection implies mutations accumulated as the result of drift, without selective advantage or disadvantage (Jeffares et al. 2015). By employing this analysis, researchers gain deeper insights into the evolutionary dynamics of viral genes, aiding in the understanding of adaptation, immune escape, and potential therapeutic targets.

6.1.3.6 Calculation of the Codon Adaptation Index (CAI)

The CAI is a key metric employed to evaluate the bias in synonymous codon usage, providing insights into the adaptation of viral genes to their host. This metric quantifies the similarity between the codon usage pattern of a specific gene and the synonymous codon frequency of a reference set, which is typically derived from the host organism. The CAI is widely used to predict potential gene expression levels; higher CAI values suggest stronger adaptation and greater preference for codons frequently used by the host. CAI values range from 0 to 1, with values closer to 1 reflecting stronger codon usage bias and greater adaptation to the host's codon preferences. Genes with higher CAI values are assumed to be better adapted to the host's translational machinery and are considered more likely to be expressed efficiently (Sharp and Li 1987). In viral genomics, the CAI helps assess the extent to which a viral gene is aligned with its host, offering valuable insights into viral-host interactions and potential pathogenicity. For the calculation of CAI, the synonymous codon usage data of *Homo sapiens*—downloaded from the Codon Usage Database (CUD)—is commonly used as the reference dataset (Nakamura et al. 2000). Viral gene sequences are compared against this reference set to calculate CAI values. The calculations are performed using CAIcal, an established tool for codon adaptation analysis (Puigbò et al. 2008). By applying CAI analysis, researchers can assess how closely viral genes mimic host codon usage, providing insights into viral evolution, expression potential, and host adaptation.

6.1.3.7 Correspondence Analysis (CoA)

Correspondence Analysis (CoA) is a multivariate statistical technique that can be applied to RSCU values, enabling the exploration of patterns and trends in codon preference across different gene sequences (Perrière and Thioulouse 2002; Suzuki et al. 2008). The analysis includes 59 codons; methionine (ATG), tryptophan (TGG), and stop codons are excluded as they do not exhibit synonymous codon usage variability. The CoA results are visualized to interpret patterns and clustering among groups of sequences, which can be categorized by genes or other characteristics. Each sequence is projected along multiple vector axes based on its RSCU profile, allowing researchers to evaluate how closely the codon usage of a particular gene aligns with that of others in the dataset. This visualization provides insights into the evolutionary forces that shape codon usage patterns, including mutational pressure and natural selection. Therefore, applying CoA to viral genetic data can reveal important aspects of viral adaptation, expression potential, and host specificity.

6.2 Case Study 1: Codon Usage Patterns and Evolutionary Dynamics of Human Papillomavirus (HPV) Oncogenes

6.2.1 Case Overview

Cervical cancer is the fourth most common cancer in women (Bray et al. 2018) and is primarily caused by persistent infection with high-risk HPV subtypes (Okunade, 2020). Over 200 HPV types have been identified (Unger and Eliav 2004), with 15 classified as high-risk due to their association with cervical cancer. The E6 and E7 proteins of high- and low-risk viruses exhibit notable similarities in their amino acid composition and structural organization; however, they differ significantly in various biochemical properties and transforming capabilities. In high-risk HPV subtypes, the products of viral transforming *E6* and *E7* genes interfere with the key tumor suppressor proteins, p53 and Rb (Crook et al. 1991; Dyson et al. 1989; Münger and Howley 2002; Wu et al. 2006). Understanding the genetic variability and codon usage patterns of these oncogenes offers new insights into the evolutionary strategies that HPV subtypes use to persist and replicate in human hosts, revealing implications for public health strategies and vaccine development (Cho et al. 2021).

6.2.2 Methodology and Data Collection

To explore the evolutionary forces shaping HPV oncogenes, sequences of the *E6* and *E7* genes from high-risk (HPV-16, 18, 31, 33, 45, 58) and low-risk (HPV-6, 11) subtypes were retrieved from the NCBI GenBank database. Sequences were filtered to maintain only those with accurate codon information and appropriate length (140–160 bp for *E6*, 90–120 bp for *E7*). For detailed analytical methods, please refer Sect. 6.1.

6.2.3 Key Findings and Interpretation

6.2.3.1 Nucleotide Composition and Codon Bias

Analysis of HPV nucleotide sequences revealed a pronounced preference for A/T-rich codons. Across both *E6* and E7 genes, adenine (A) was identified as the most common nucleotide, with its composition ranging between 32.46% and 38.18% in the *E6* gene and 28.63% and 32.39% in the *E7* gene. Further evidence of this AT-rich composition was observed at the third codon position (AT3), where A-ended codons were overrepresented. The analysis of codon usage bias in HPV subtypes revealed distinct patterns between the *E6* and *E7* genes, with varying degrees of bias across subtypes. To quantify this bias, the ENC was calculated for both genes. In the *E6* gene, HPV-58 exhibited the strongest bias with the lowest ENC value (40.18), indicating limited use of synonymous codons; HPV-6 had the highest ENC value (58.98), reflecting a more relaxed codon usage. Conversely, for the *E7* gene, HPV-6 displayed the lowest ENC value (44.81), suggesting higher bias, whereas HPV-58 had the highest ENC value (60.98), indicating more diverse codon usage. Interestingly, high-risk subtypes (e.g., HPV-16, HPV-31, and HPV-58) demonstrated significant variability between the two genes. For example, while HPV-16, HPV-31, and HPV-58 had high ENC values (>50) in the *E7* gene, they exhibited more constrained codon usage patterns in the *E6* gene. This suggests that the two oncogenes, despite sharing similar roles in viral oncogenesis, may have evolved distinct codon preferences to optimize their functions within the viral life cycle. Low-risk subtypes such as HPV-6 and HPV-11 generally displayed a more relaxed codon usage bias, with higher ENC values for both genes indicating a broader utilization of synonymous codons. This difference in codon usage between high- and low-risk subtypes may reflect divergent evolutionary strategies, with high-risk subtypes potentially fine-tuning codon usage to enhance viral protein translation efficiency and promote oncogenesis. The contrasting codon usage patterns suggest that HPV subtypes have evolved different levels of codon bias tailored to their replication and survival needs. This variation in codon bias might play a role in the tumorigenic potential of the virus; more biased codon usage in high-risk subtypes may contribute to efficient viral protein synthesis, enhancing viral persistence and the likelihood of tumor development in host cells.

6.2.3.2 Evolutionary Forces: Mutational Pressure vs. Selection Pressure

Neutrality plots comparing GC content across codon positions (GC12 vs. GC3) highlighted the complex interaction between mutational pressure and natural selection. For most HPV subtypes, natural selection was the dominant force; this was evidenced by dN/dS ratios below 1, indicating purifying selection. However, specific subtypes, such as HPV-33 (*E6*) and HPV-58 (*E7*), showed higher mutational pressure, reflected in their codon usage variation. The usage patterns of these subtypes deviated from the expected neutrality, indicating that mutation played a more significant role in their evolutionary history. For example, the neutrality plot revealed that in the *E6* gene, HPV-33 codon usage bias had a relative neutrality (mutation pressure) of 30.95%, reflecting a notable influence of mutation. Conversely, most subtypes exhibited values indicating that natural selection, rather than mutation, governed codon usage. This suggests that the virus maintains a fine balance between mutational variability and functional protein constraints.

6.2.3.3 Codon Adaptation and Host Interactions

The CAI was used to assess how well the codon usage of HPV subtypes aligns with that of the human host. CAI values ranged between 0.71 and 0.77 for the *E6* gene and 0.71 and 0.75 for the *E7* gene, with HPV-16 showing the highest values for both. These high values suggest that HPV-16 has optimized its codon usage to align with the codon preferences of human epithelial cells, facilitating viral persistence and oncogenesis. These findings imply that codon adaptation plays a crucial role in the virus's ability to persist within the host by avoiding translational bottlenecks and immune surveillance mechanisms. The preference for codons aligned with the host's translational machinery could also facilitate rapid viral replication and protein synthesis, contributing to pathogenicity. Elevated CAI values suggest that high-risk subtypes, such as HPV-16, have an evolutionary advantage in establishing persistent infections and promoting oncogenesis. The alignment of codon usage patterns between the virus and host highlights the intricate co-evolution of HPV with human cells. It underscores the importance of codon optimization in viral evolution and suggests potential avenues for therapeutic interventions, such as targeting specific codon usage patterns to disrupt viral replication. Understanding the CAI-based adaptation of HPV subtypes could provide the basis for designing codon-optimized vaccines or gene therapies that can better target the virus and prevent its oncogenic transformation.

6.2.4 Comparative Insights

The analysis of codon usage patterns in HPV subtypes revealed their commonalities and unique distinctions, especially between high- and low-risk subtypes. High-risk subtypes exhibited evidence of stronger selection pressures, reflected in more biased codon usage. This bias suggests a selective strategy to optimize protein synthesis, which may contribute to the higher oncogenic potential of these subtypes. These high-risk variants, such as HPV-16, exhibited greater alignment with human codon usage, suggesting an adaptation that enhances translation efficiency and persistence within human epithelial cells. In contrast, low-risk subtypes, like HPV-6 and HPV-11, displayed higher ENC values (e.g., 58.98 for *E6* in HPV-6), indicating a more relaxed codon usage bias. These subtypes also demonstrated higher variability in their codon preferences, reflecting a less stringent evolutionary adaptation consistent with their lower pathogenicity. Notably, patterns differed between the *E6* and *E7* genes within the same subtype, highlighting how evolutionary forces may shape specific genes differently. For instance, HPV-6 exhibited tighter codon usage bias in *E7* (ENC: 44.81) than in *E6*, while HPV-58 showed the reverse trend. These findings underscore that the evolution of codon usage bias within HPV subtypes is multifaceted, driven by both selective pressures and adaptation to the host environment. These factors help shape their pathogenicity and ability to evade immune detection.

6.2.5 Applications and Public Health Implications

The analysis of codon usage patterns across HPV subtypes offers essential insights with far-reaching public health implications, particularly for vaccine development and disease prevention efforts. Identifying the preferred codon usage patterns of high-risk subtypes enables the development of codon-optimized vaccines that enhance antigen expression and elicit stronger immune responses. Codon modification, especially in the E6 and E7 proteins which play crucial roles in HPV-related oncogenesis, has already demonstrated success in experimental vaccines by overcoming translational blockages and improving antigen yields (Lin et al., 2006; Qin et al., 2021). Such optimization will be instrumental in developing the next generation of vaccines, specifically targeting high-risk HPV subtypes to more effectively prevent associated cancers. Furthermore, monitoring changes in codon usage patterns provides insights into the ongoing evolutionary dynamics of HPV. Viral adaptations, such as changes in tissue tropism or immune escape mechanisms, can be detected early by tracking codon shifts, guiding surveillance programs and public health interventions. This approach helps anticipate emerging vaccine escape variants or resistance to therapeutic interventions, enabling proactive mitigation strategies. Aligning immunization efforts with the evolutionary dynamics of HPV subtypes will strengthen vaccination campaigns and improve public health outcomes, reducing the burden of HPV-related diseases over the long term.

6.2.6 Conclusion and Future Directions

This case study emphasizes the importance of codon usage bias and evolutionary forces in shaping the pathogenicity and persistence of HPV subtypes. While natural selection is the primary driver influencing codon usage, mutational pressure also contributes to codon variability, as observed through neutrality plots and dN/dS analyses. Understanding the interplay between these evolutionary forces offers critical insights into viral persistence, immune evasion, and tissue tropism.

Future research should further explore the interactions between codon preferences and host immune responses. Unraveling how HPV proteins evade immune detection through optimized codon usage could shed light on mechanisms of viral persistence. Moreover, therapeutic interventions, including DNA vaccines that leverage codon optimization technologies, represent a promising avenue for enhancing treatment outcomes. The presented findings also emphasize the importance of tracking codon usage trends as new HPV subtypes continue to emerge. Ongoing research into these genetic adaptations will be critical for the timely identification of novel clinical challenges and the development of targeted solutions. Insights from codon usage analysis can, additionally, inform therapeutic strategies aimed at suppressing viral replication or preventing the development of HPV-related malignancies. This knowledge will not only guide future research but also support policy decisions and global health initiatives focused on mitigating HPV-related health burdens.

6.3 Case Study 2: Codon Usage Profiles of Large Tumor Antigen (LT-Ag) Genes in Polyomaviruses Across Various Host Species

6.3.1 Case Overview

PyVs are double-stranded DNA viruses that infect a diverse array of hosts, from mammals and birds to fish (ICTV, https://ictv.global/report_9th/dsDNA/Polyomaviridae). Their evolutionary dynamics reveal significant differences in host adaptation, genetic variability, and pathogenicity. Among their genes, LT-Ag is particularly crucial, as it plays a direct role in tumorigenesis through interactions with host proteins such as p53 and Rb (Ahuja et al. 2005; Shuda et al. 2008). This case study focuses on the codon usage patterns of LT-Ag across different host species, comparing genetic and evolutionary traits and exploring how these differences influence host range, pathogenicity, and virus-host interactions (Cho et al. 2019).

6.3.2 Methodology and Data Collection

Codon usage analysis was conducted using the CDS of *LT-Ag* from 86 PyV species. These were classified into five host categories: mammals (Group M), non-human primates (Group P), humans (Group H), birds (Group A), and fish (Group F). Reference sequences were obtained from GenBank, and the domain sequences of LT-Ag, including the DnaJ and helicase domains, were extracted using PROSITE tools. For detailed analytical methods, please refer Sect. 6.1.

6.3.3 Key Findings and Interpretation

6.3.3.1 Phylogenetic Relationships and Evolutionary Insights

Phylogenetic analyses of the *LT-Ag* gene and its DnaJ and helicase domains revealed distinct evolutionary clusters. Alphapolyomavirus and Betapolyomavirus were grouped separately, with notable exceptions such as bat PyV 2c and bat PyV 4a, which displayed divergent patterns. In contrast, Gammapolyomavirus formed an independent cluster. Deltapolyomavirus and some unclassified PyVs either clustered together or appeared isolated. When analyzed by the host, Group H, P, and M PyVs clustered closely, while Group A and F PyVs formed distinct, independent groups. These findings highlight host-specific evolution and adaptive strategies. Avian PyVs displayed higher sequence variability, correlating with their broader host range and greater codon diversity.

6.3.3.2 Compositional Differences and Codon Bias

LT-Ag sequences exhibited host-specific codon biases. Group A and F viruses exhibited higher GC and GC3 values (>45). In contrast, Groups H, M and P exhibited higher AT and AT3 values (>60); this was particularly elevated in Group H (>70). This AT-rich composition aligns with previous reports suggesting host-specific genome adaptations. ENC analysis revealed that avian and fish PyVs exhibited greater codon diversity (ENC > 50), while mammalian PyVs (Groups H, P, and M) demonstrated more biased codon usage (ENC < 45), reflecting evolutionary constraints associated with host adaptation and viral persistence.

6.3.3.3 Host-Specific Codon Usage Patterns

Distinct codon usage preferences were observed across hosts. For example, mammalian viruses (Groups H, P, and M) favored TTT (phe), TTA (leu), ATT (ile), and GCT (ala), differing from Groups A and F. The codon AGA (arg) exhibited the most pronounced variation in usage preference across the groups; its mean RSCU values were calculated as 1.55 for Group A, 2.07 for Group F, 4.40 for Group H, 3.90 for Group M, and 4.28 for Group P. Avian PyVs, infecting penguins, geese, and other birds, demonstrated broad host range adaptability, aligning with their high codon diversity. Conversely, mammalian PyVs showed more host-specific codon usage, indicating more conservative evolutionary strategies.

This pattern reflects the evolutionary pressures that shape codon preferences, where more specialized codon usage in mammals ensures higher translation efficiency within specific hosts, potentially affecting viral replication and pathogenicity.

6.3.3.4 Selection Pressure

Selection pressure analyses using dN/dS ratios revealed both the DnaJ and helicase domains to be under negative selection pressure (dN/dS < 1) in all groups. This suggests that these functional regions are conserved to maintain critical viral functions. However, in some comparisons, human PyVs exhibited positive selection (dN/dS > 1), indicating ongoing evolutionary changes likely driven by host immune interactions. Neutrality analysis showed varying contributions of mutational pressure across groups, with Group F exhibiting the highest mutation-driven evolution (59.57%) and Group A conversely showing strong natural selection influence (83.46%). These findings emphasize that codon usage patterns are shaped by both mutational forces and adaptive selection to enhance viral survival and host interaction across species.

6.3.4 Comparative Insights

Comparative analysis revealed significant differences in codon usage patterns, evolutionary constraints, and host-specific adaptation strategies between mammalian, avian, and fish PyVs. Mammalian PyVs exhibited lower codon diversity and ENC values, indicating constrained evolution driven by purifying selection. These viruses tend to have narrower host specificity and rely on stable host-virus relationships, rarely causing acute disease in their natural hosts unless the immune system is compromised. For example, SV40 induces tumors only in nonpermissive rodents (Todaro and Green 1966); this showcases the host-restricted nature that is common to mammalian PyVs. In contrast, avian PyVs showed greater codon diversity and higher GC content, consistent with their broader host ranges and greater potential for cross-species transmission. This diversity reflects their ability to infect various bird species, such as geese and penguins, and cause severe diseases like hemorrhagic nephritis and BFDV-induced immunosuppression (Bernier et al. 1981; Bozeman et al. 1981; Johne and Müller 1998; Johne et al. 2006; Guerin et al. 2000). The high pathogenicity of avian PyVs, linked to efficient cell destruction and necrosis, further underscores the evolutionary differences in host-pathogen interactions. Fish PyVs, though evolutionarily distant from their mammalian and avian counterparts, demonstrated the highest mutation pressure, reflected in their neutrality plots. This suggests that fish PyVs are subject to distinct evolutionary forces, possibly due to the environmental variability in aquatic ecosystems driving rapid genetic shifts. These findings highlight how codon usage diversity and host adaptation shape the evolutionary trajectories of PyVs across different taxa.

6.3.5 Applications and Public Health Implications

The study of LT-Ag codon usage patterns offers valuable insights for public health strategy and vaccine development. Codon bias analysis can predict emerging cross-species transmissions, as seen in avian PyVs with high codon diversity, which indicates their potential for host-switching events. This emphasizes the importance of genomic surveillance to monitor codon shifts, offering early warnings of potential outbreaks or new pathogenic variants. These findings also play a crucial role in vaccine development. Codon optimization of viral proteins, such as LT-Ag, can improve antigen expression and enhance immune responses, especially for high-risk PyV strains like MCPyV, which is linked to Merkel cell carcinoma. Optimized vaccines can also be tailored to elicit stronger immune reactions in immunocompromised individuals—a population particularly vulnerable to viral reactivation. Ultimately, a better understanding of host-specific codon preferences will allow for the development of targeted vaccines and therapies, improving the management of PyV-related diseases across diverse host species.

6.3.6 Conclusion and Future Directions

This study emphasizes the importance of codon usage bias and evolutionary dynamics in shaping the pathogenicity and adaptability of PyVs. Mammalian PyVs exhibit conservative evolution, driven primarily by purifying selection, while avian and fish PyVs show greater codon diversity and adaptability, contributing to their broader host ranges and higher virulence. These insights into codon usage patterns provide a framework for understanding viral evolution, immune evasion mechanisms, and host interactions. Future research should focus on how codon preferences influence viral persistence and immune responses in different hosts. Investigating the interplay between mutational pressure and selection forces will offer deeper insights into host-virus co-evolution. Monitoring codon usage evolution will remain crucial for anticipating emerging viral threats and guiding vaccine strategies. Moreover, expanding studies to include newly identified PyVs will further enhance our understanding of their ecological impact, improving therapeutic interventions and public health management.

References

Ahuja D, Sáenz-Robles MT, Pipas JM. SV40 large T antigen targets multiple cellular pathways to elicit cellular transformation. Oncogene. 2005;24:7729–45.

Bernier G, Morin M, Marsolais G. A generalized inclusion body disease in the budgerigar (*Melopsittacus undulatus*) caused by a papovavirus-like agent. Avian Dis. 1981;25:1083–92.

Bozeman LH, Davis RB, Gaudry D, Lukert PD, Fletcher OJ, Dykstra MJ. Characterization of a papovavirus isolated from fledgling budgerigars. Avian Dis. 1981;25:972–80.

Bray F, Ferlay J, Soerjomataram I, Siegel RL, Torre LA, Jemal A. Global cancer statistics 2018: GLOBOCAN estimates of incidence and mortality worldwide for 36 cancers in 185 countries. CA Cancer J Clin. 2018;68:394–424.

Chenna R, Sugawara H, Koike T, Lopez R, Gibson TJ, Higgins DG, Thompson JD. Multiple sequence alignment with the Clustal series of programs. Nucleic Acids Res. 2003;31:3497–500.

Cho M, Kim H, Son HS. Codon usage patterns of LT-Ag genes in polyomaviruses from different host species. Virol J. 2019;16:1–24.

Cho M, Kim H, Je M, Son HS. Analysis of codon usage patterns in the human papillomavirus oncogenes. Curr Bioinforma. 2021;16:13–23.

Comeron JM, Aguadé M. An evaluation of measures of synonymous codon usage bias. J Mol Evol. 1998;47:268–74.

Crook T, Tidy JA, Vousden KH. Degradation of p53 can be targeted by HPV E6 sequences distinct from those required for p53 binding and trans-activation. Cell. 1991;67:547–56.

Dyson N, Howley PM, Münger K, Harlow E. The human papillomavirus-16 E7 oncoprotein is able to bind to the retinoblastoma gene product. Science. 1989;243:934–7.

Edgar RC. MUSCLE: multiple sequence alignment with high accuracy and high throughput. Nucleic Acids Res. 2004;32:1792–7.

Felsenstein J. Confidence limits on phylogenies: an approach using the bootstrap. Evolution. 1985;39:783–91.

Fu YX, Li WH. Statistical tests of neutrality of mutations. Genetics. 1993;133:693–709.

Guerin JL, Gelfi J, Dubois L, Vuillaume A, Boucraut-Baralon C, Pingret JL. A novel polyomavirus (goose hemorrhagic polyomavirus) is the agent of hemorrhagic nephritis enteritis of geese. J Virol. 2000;74:4523–9.

Guindon S, Dufayard JF, Lefort V, Anisimova M, Hordijk W, Gascuel O. New algorithms and methods to estimate maximum-likelihood phylogenies: assessing the performance of PhyML 3.0. Syst Biol. 2010;59:307–21.

Huelsenbeck JP, Ronquist F. MRBAYES: Bayesian inference of phylogenetic trees. Bioinf. 2001;17:754–5.

ICTV (International Committee on Taxonomy of Viruses) (n.d.). *Polyomaviridae*. https://ictvglobal/report_9th/dsDNA/Polyomaviridae.

Jeffares DC, Tomiczek B, Sojo V, dos Reis M. A beginner's guide to estimating the non-synonymous to synonymous rate ratio of all protein-coding genes in a genome. Parasite genomics protocols. Method Mol Biol. 2015;1201:65–90.

Johne R, Müller H. Avian polyomavirus in wild birds: genome analysis of isolates from Falconiformes and Psittaciformes. Arch Virol. 1998;143:1501–12.

Johne R, Wittig W, Fernández-de-Luco D, Höfle U, Müller H. Characterization of two novel polyomaviruses of birds by using multiply primed rolling-circle amplification of their genomes. J Virol. 2006;80:3523–31.

Katoh K, Kuma K, Toh H, Miyata T. MAFFT version 5: improvement in accuracy of multiple sequence alignment. Nucleic Acids Res. 2005;33:511–8.

Kryazhimskiy S, Plotkin JB. The population genetics of dN/dS. PLoS Genet. 2008;4:e1000304.

Kumar S, Stecher G, Tamura K. MEGA7: molecular evolutionary genetics analysis version 7.0 for bigger datasets. Mol Biol Evol. 2016;33:1870–4.

Li SS. Phylogenetic tree construction using Markov Chain Monte Carlo [PhD dissertation]. Columbus: Ohio State University; 1996.

Librado P, Rozas J, Dna SP. v5: a software for comprehensive analysis of DNA polymorphism data. Bioinf. 2009;25:1451–2.

Lin C-T, Tsai Y-C, He L, Calizo R, Chou H-H, Chang T-C, Soong Y-K, Hung C-F, Lai C-H. A DNA vaccine encoding a codon-optimized human papillomavirus type 16 E6 gene enhances CTL response and anti-tumor activity. J Biomed Sci. 2006;13:481–8.

Mau B. Bayesian phylogenetic inference via Markov Monte Carlo methods. Ph.D. dissertation, University of Wisconsin; 1996.

Mau B, Newton MA. Phylogenetic inference for binary data on dendrograms using Markov chain Monte Carlo. J Comput Graph Stat. 1997;6:122–31.

Mau B, Newton MA, Larget B. Bayesian phylogenetic inference via Markov chain Monte Carlo methods. Biometrics. 1999;55:1–12.

Münger K, Howley PM. Human papillomavirus immortalization and transformation functions. Virus Res. 2002;89:213–28.

Nakamura Y, Gojobori T, Ikemura T. Codon usage tabulated from international DNA sequence databases: status for the year 2000. Nucleic Acids Res. 2000;28:292.

Nei M, Gojobori T. Simple methods for estimating the numbers of synonymous and nonsynonymous nucleotide substitutions. Mol Biol Evol. 1986;3:418–26.

Okunade KS. Human papillomavirus and cervical cancer. J Obstet Gynaecol. 2020;40:602–8.

Perrière G, Thioulouse J. Use and misuse of correspondence analysis in codon usage studies. Nucleic Acids Res. 2002;30:4548–55.

Puigbò P, Bravo IG, Garcia-Vallve S. CAIcal: a combined set of tools to assess codon usage adaptation. Biol Direct. 2008;3:38.

Qin F, Xia F, Chen H, Cui B, Feng Y, Zhang P, Chen J, Luo M. A guide to nucleic acid vaccines in the prevention and treatment of infectious diseases and cancers: from basic principles to current applications. Front Cell Dev Biol. 2021;9:633776.

Rannala B, Yang Z. Probability distribution of molecular evolutionary trees: a new method of phylogenetic inference. J Mol Evol. 1996;43:304–11.

Saitou N, Nei M. The neighbor-joining method: a new method for reconstructing phylogenetic trees. Mol Biol Evol. 1987;4:406–25.

Sharp PM, Li WH. An evolutionary perspective on synonymous codon usage in unicellular organisms. J Mol Evol. 1986a;24:28–38.

Sharp PM, Li WH. Codon usage in regulatory genes in *Escherichia coli* does not reflect selection for 'rare' codons. Nucleic Acids Res. 1986b;14:7737–49.

Sharp PM, Li WH. The codon adaptation index—a measure of directional synonymous codon usage bias, and its potential applications. Nucleic Acids Res. 1987;15:1281–95.

Shuda M, Feng H, Kwun HJ, Rosen ST, Gjoerup O, Moore PS, Chang Y. T antigen mutations are a human tumor-specific signature for Merkel cell polyomavirus. Proc Natl Acad Sci. 2008;105:16272–7.

Sueoka N. Directional mutation pressure and neutral molecular evolution. Proc Natl Acad Sci. 1988;85:2653–7.

Sueoka N. Intrastrand parity rules of DNA base composition and usage biases of synonymous codons. J Mol Evol. 1995;40:318–25.

Sueoka N. Translation-coupled violation of Parity Rule 2 in human genes is not the cause of heterogeneity of the DNA G + C content of third codon position. Gene. 1999;238:53–8.

Suzuki H, Brown CJ, Forney LJ, Top EM. Comparison of correspondence analysis methods for synonymous codon usage in bacteria. DNA Res. 2008;15:357–65.

Tajima F. Statistical method for testing the neutral mutation hypothesis by DNA polymorphism. Genetics. 1989;123:585–95.

Tamura K, Nei M. Estimation of the number of nucleotide substitutions in the control region of mitochondrial DNA in humans and chimpanzees. Mol Biol Evol. 1993;10:512–26.

Todaro GJ, Green H. High frequency of SV40 transformation of mouse cell line 3T3. Virology. 1966;28:756–9.

Tsompana M, Abad J, Purugganan M, Moyer JW. The molecular population genetics of the tomato spotted wilt virus (TSWV) genome. Mol Ecol. 2005;14:53–66.

Unger ER, Eliav B. Human papillomavirus and cervical cancer. Emerg Infect Dis. 2004;10:2031–2.

Wong EH, Smith DK, Rabadan R, Peiris M, Poon LL. Codon usage bias and the evolution of influenza A viruses. BMC Evol Biol. 2010;10:253.

Wright F. The 'effective number of codons' used in a gene. Gene. 1990;87:23–9.

Wu Y, Chen Y, Li L, Yu G, He Y, Zhang Y. Analysis of mutations in the E6/E7 oncogenes and L1 gene of human papillomavirus 16 cervical cancer isolates from China. J Gen Virol. 2006;87:1181–8.

Yang Z, Rannala B. Bayesian phylogenetic inference using DNA sequences: a Markov Chain Monte Carlo method. Mol Biol Evol. 1997;14:717–24.

Yang Z, Bielawski JP. Statistical methods for detecting molecular adaptation. Trends Ecol Evol. 2000;15:496–503.

7 Prediction of Host Tropism and Cross-Species Infection of Viruses

Viruses display a remarkable spectrum of host specificities, ranging from those restricted to a single species to others capable of infecting diverse hosts across taxa. Host tropism, defined as the ability of a virus to infect specific host species, arises from intricate interactions between viral proteins and the host's cellular machinery. This selectivity is central to understanding viral evolution, cross-species transmission, and the emergence of new viral variants. Critical factors, such as viral receptor binding efficiency, replication capacity, and immune evasion strategies, are governed by molecular interactions at the protein level and ultimately determine the host range of a virus. These factors significantly influence a virus's potential to trigger epidemics or pandemics, emphasizing the importance of understanding host tropism in virology and public health. In recent years, machine learning has become an essential tool for studying host-virus interactions, offering robust methodologies for identifying molecular markers that govern host specificity and cross-species infection risks. Machine learning models have been instrumental in uncovering how specific amino acid substitutions in viral proteins can expand or constrain the host range of a virus, such as those in spike protein receptor-binding domains (RBDs) of coronaviruses or hemagglutinin (HA) in influenza viruses. These predictive approaches have profound implications for zoonotic surveillance, enabling the early detection of host-switching viruses like avian influenza or SARS-CoV-2. By identifying key molecular features, machine learning facilitates the assessment of cross-species infection risks and informs strategies for mitigating emerging viral threats.

In this chapter, we will examine the primary analytical methods employed to predict host tropism and cross-species infection propensities, as outlined in the sections on analytical methods. These include feature extraction techniques, which capture critical molecular properties of viral proteins and host receptors; machine learning algorithms such as RF and linear discriminant analysis (LDA), which enable robust classification and predictive modeling; and performance evaluation metrics, which ensure the reliability and interpretability of machine learning-driven predictions. Together, these methods form a comprehensive computational framework for studying viral host adaptation. To demonstrate these approaches in practical contexts, we present two case studies. The first case study uses machine learning to identify key molecular features associated with host specificity to explore host tropism determinants in influenza viruses (Kwon et al. 2020). The second case study investigates the cross-species infection propensities of viruses by analyzing sequence similarity and evolutionary distance in host receptor proteins (Cho and Son 2019). Through these case studies, we highlight the power of computational tools and machine learning in elucidating the

H. S. Son, M. Cho, *Viroinformatics and Computational Virology in Public Health*,
https://doi.org/10.1007/978-981-96-5757-5_7

dynamics of viral host adaptation and cross-species transmission. These insights extend beyond theoretical research, providing actionable data to enhance surveillance, guide vaccine development, and strengthen public health preparedness. By integrating advanced computational techniques with virology research, we can better understand the mechanisms driving viral transmission and evolution, ultimately informing more effective strategies to address emerging infectious diseases.

7.1 Overview of Analytical Methods in Host Tropism and Cross-Species Infection of Viruses

7.1.1 Feature Extraction

Feature extraction is the foundational step in the development of predictive models for viral host tropism and cross-species infection propensities, as it involves the identification and quantitative representation of molecular attributes that influence host-virus interactions. In host tropism analysis, viral proteins—such as the HA and neuraminidase (NA) proteins in influenza viruses—are often the focus, as these proteins play key roles in host recognition and binding. Key molecular features include amino acid composition, hydrophobicity, van der Waals volume, charge, polarity, and solvent accessibility. These physicochemical properties reflect aspects of viral protein interactions with host cellular receptors, which are central to understanding host specificity and adaptability.

To accurately quantify these features, online servers designed for bioinformatics analysis like PROFEAT (Li et al. 2006) and iFEATURE (Chen et al. 2018) can be employed to calculate composition (C), transition (T), and distribution (D) values across amino acid sequences (Dubchak et al. 1995, 1997). The C-T-D metrics reflect the physicochemical characteristics of amino acid sequences, providing a structured, quantitative input for machine learning models. By transforming sequence data into these feature vectors, the model can capture critical determinants of host specificity. PROFEAT and iFEATURE also offer a range of methods for feature extraction, allowing users to analyze a broader set of molecular properties such as polarity, charge, and solvent accessibility, essential for understanding protein interactions with host receptors. This enables the transformation of protein sequence data into feature vectors that effectively encode host-specific determinants.

For cross-species infection, feature extraction focuses on host receptor proteins, analyzing sequence similarities and evolutionary distances between the receptors of potential host species. MSA and scoring matrices, for instance, are applied to compute absolute pairwise distances, relative distances, and total sequence similarity—core predictive variables for assessing cross-species infection risk. This integrated feature extraction approach enhances a model's capacity to identify infection propensity and host range specificity, which are essential for predicting zoonotic potential.

7.1.2 Machine Learning Analysis

Various machine learning algorithms can be applied to classify and predict host tropism and cross-species infection risk, each suited to different types of analysis and data characteristics. For host tropism research, RF (Breiman 2001), kNN (Fix and Hodges 1951), and NB (Bayes 1763) approaches are appropriate for classification tasks. These algorithms use physicochemical properties extracted from viral protein sequences to distinguish between different host classes, such as avian, human, and swine, in the case of influenza viruses. For instance, the RF ensemble approach, which combines the predictions of multiple decision trees, can be particularly effective, producing stable and accurate classifications by mitigating overfitting. Furthermore, feature selection methods—such as correlation-based selection or ranking-based attribute selection (Hall 1999; Roffo and Melzi 2016)—can be used to refine models by prioritizing amino acid properties with the greatest influence on host classification, focusing the model on key variables and improving interpretability.

For cross-species infection predictions, LDA (Fisher 1936) can be particularly effective due to its ability to model the linear separation between groups (infectious vs. non-infectious) based on sequence similarity measures. LDA identifies optimal linear combinations of variables—such as evolutionary distances and sequence similarities—to maximize the separation between groups, providing a discriminant score that quantifies infection risk. This score, derived from the model's discriminant function, effectively assigns host pairs to risk categories, with higher scores indicating a greater potential for infection. LDA's ability to maximize group separation through linear discrimination makes it ideal for cross-species infection studies, where subtle variations in receptor similarity significantly impact risk classification.

7.1.3 Performance Evaluation and Optimization

Evaluating model performance is crucial for ensuring the reliability and accuracy of predictions. Standard metrics include accuracy, precision, sensitivity, specificity, MCC (Matthews 1975), and the area under the curve (AUC). Accuracy is commonly cross-validated through techniques that enhance reliability by assessing the model across multiple training and testing subsets. RF models can achieve high accuracy and stability due to their ensemble structure, integrating multiple decision trees to reduce prediction variance.

LDA model performance is also commonly evaluated through sensitivity and specificity measures, with a focus on the discriminant model's ability to accurately distinguish between different classes. Leave-one-out cross-validation (LOOCV) is widely used to validate LDA-based models, improving robustness by iteratively training the model on all data points except one and testing on the excluded point (Hastie et al. 2009). Model optimization includes refining discriminant coefficients or using feature selection to prioritize the most predictive variables effectively. High specificity and sensitivity values (approaching 100%) demonstrate the discriminant model's precision in classification tasks, which is essential for applications ranging from epidemiological predictions to targeted surveillance.

In summary, combining feature extraction, machine learning, and rigorous performance evaluation can optimize the predictive capability of models, allowing for reliable and insightful predictions. These optimized models provide essential support for data-driven strategies in fields like public health, enabling proactive measures and informing risk management decisions across various domains.

7.2 Case Study 1: Deciphering Host Tropism of Influenza Viruses Using Machine Learning

7.2.1 Case Overview

This case study investigates the determinants of host tropism in influenza viruses, with a particular focus on molecular markers and physicochemical properties that predict host specificity (Kwon et al. 2020). This research uses machine learning models to identify the key physicochemical properties of amino acids within viral proteins that influence host prediction, ultimately facilitating risk assessment and public health surveillance. By examining the amino acid sequences of HA, NA, polymerase (PA, PB1, and PB2), and nucleoprotein (NP), which act as host determinants for the influenza virus, this study underscores the role of certain critical proteins in driving host specificity and enabling cross-species transmission. HA is highlighted for its role in receptor binding and viral entry, both central to host adaptation and zoonotic potential.

7.2.2 Methodology and Data Analysis

The data for this study comprised over 109,451 protein sequences from six viral proteins (HA, NA, PB1, PB2, PA, PB1, PB2, and NP) associated with influenza host tropism. These sequences were sourced from avian, human, and swine influenza strains. To focus on the molecular characteristics of viral proteins, amino acid sequences were processed and filtered, removing incomplete or redundant sequences. Each protein sequence was converted into feature vectors to quantify 24 physicochemical properties, such as hydrophobicity, van der Waals volume, and charge, using the online PROFEAT server (Li et al. 2006). CTD values and amino acid composition (AAC) were computed to capture the primary physicochemical properties essential for machine learning models (Dubchak et al. 1995, 1997). The analysis applied three machine learning algorithms—RF (Breiman 2001), kNN (Fix and Hodges 1951), and NB (Bayes 1763)—to classify viral sequences according to host types. RF, known for its robustness in handling large datasets, was optimized using ranking-based feature selection, allowing the model to prioritize the most significant features and exclude less relevant or noisy data points (Hall 1999; Roffo and Melzi 2016). This step optimized the feature set, reducing computational load while preserving predictive power. For each algorithm, ten-fold cross-validation was employed to evaluate model stability across protein types and validate classification accuracy (Hastie et al. 2009). Statistical analyses, including ANOVA (Fisher 1970) and Tukey's post-hoc tests (Tukey 1949), were conducted to confirm the significance of the selected physicochemical features across host groups. Finally, ranking-based feature selection helped isolate highly influential features, enabling the model to identify essential physicochemical properties that most effectively distinguish host species.

7.2.3 Key Findings and Interpretation

7.2.3.1 Model Performance and Protein-Specific Accuracy

Among the machine learning algorithms tested, RF was the best performing, achieving a 93.62% accuracy and 0.98 AUC across all proteins. Specifically, the HA protein achieved the highest classification accuracy, at 96.19% across host species. The high performance of HA highlights its pivotal role in receptor binding and viral entry, which has a direct impact on the host range and adaptability of the virus (Schneider-Schaulies 2000; Hueffer et al. 2003; Imai and Kawaoka 2012). The NA protein also showed high accuracy at 96.00%, which aligns with its function in viral release through the cleavage of sialic acid receptors. The sialidase activity of NA also facilitates viral access of cells, catalyzing the cleavage of sialic acids presented by decoy receptors, such as mucins (Kesimer et al. 2009; McAuley et al. 2017). This dual functionality underscores the critical role of NA in the viral lifecycle and host adaptation, contributing to both viral release and entry (Cohen et al. 2013; McAuley et al. 2019). In polymerase proteins, PB2 exhibited a classification accuracy of 93.55%, highlighting its importance in viral replication efficiency and host adaptation across different hosts (Tarendeau et al. 2008). Roles of such proteins demonstrate the complexity of viral tropism beyond surface proteins like HA. Similarly, PA and PB1 also contributed significantly to host classification, with PA achieving an accuracy of 91.78%, and PB1 showing an accuracy of 91.32%. These findings underscore the critical roles of polymerase proteins in shaping host specificity and their combined influence on the evolutionary dynamics of viral adaptation (Bussey et al. 2011; Mehle et al. 2012). The consistently high AUC values for RF models underscore the reliability of this algorithm in predicting host specificity from viral sequence data.

7.2.3.2 Key Physicochemical Properties

These results identify critical physicochemical properties influencing host classification, providing insight into the molecular basis of viral adaptation. For HA protein, van der Waals volume, polarizability, and charge emerged as the key physicochemical factors affecting classification. These findings align with the known interaction between HA and host cell receptors, in which electrostatic charge and van der Waals forces play crucial roles in determining binding efficiency and specificity (Xu et al. 2009). Similarly, the polarity of NA emerged as the most significant predictor for host specificity, likely due to its role in cleaving sialic acid receptors, facilitating viral release (Kesimer et al. 2009; Cohen et al. 2013; McAuley et al. 2017, 2019). The analysis also showed solvent accessibility and electric charge to be primary factors for the PB1 and PB2 proteins, likely due to their roles in RNA synthesis and interactions with host cell machinery (Tarendeau et al. 2008; Bussey et al. 2011; Mehle et al. 2012). These findings highlight the distinct physicochemical attributes of viral proteins that play key roles in driving host adaptation and determining host specificity.

7.2.3.3 Host-Specific Markers and Receptor-Binding Domain Analysis

A detailed analysis of the HA RBD highlights the roles of specific amino acid residues in host specificity and cross-species transmission. Notably, positions 226 and 228 within the HA RBD have been identified as key determinants of host tropism. Mutations at position 226 significantly influence cell tropism and replication efficiency in human airway epithelial cells, enhancing the virus's ability to infect human hosts. Similarly, alterations at position 228 impact receptor binding preferences, further shaping host specificity. These findings demonstrate the varying roles of residues in mediating viral adaptation and underscore their potential importance in facilitating cross-species transmission of influenza A viruses (Vines et al. 1998; Wan and Perez 2007; Bateman et al. 2008).

7.2.3.4 Feature Reduction and Machine Learning Optimization

Feature reduction using RF models with ranking-based feature selection enhanced classification accuracy while simultaneously reducing computational demands, by focusing on the most relevant amino acid properties. The model's accuracy was maintained even with a reduced feature set—using as few as 10 features for NP and 75 features for PA—without compromising classification effectiveness. For example, RF models employing only the top 25 ranked features consistently achieved accuracy exceeding 90% across all six proteins, validating the robustness of the optimization process. This streamlined approach confirms that a carefully selected subset of key features can capture the critical molecular patterns necessary for accurate host classification. Moreover, the feature selection methodology underscores the importance of prioritizing functionally significant properties, ensuring reliable predictions even with minimal datasets. By balancing feature relevance with computational efficiency, this optimized model is well-suited for large-scale viral surveillance, facilitating rapid analyses and supporting timely public health interventions.

7.2.4 Comparative Insights and Implications for Viral Dynamics and Surveillance

This study expands on previous research by Allen et al. and Hu, both of whom explored host specificity markers in influenza viruses using machine learning. Allen et al.'s work focused on specific amino acid markers using linear support vector machines (LSVM) to explore how mutations affect host specificity and infection severity (Allen et al. 2009). Hu further advanced these efforts by applying RF models to identify significant markers across multiple influenza proteins, including HA, NA, polymerases, NP, matrix proteins, and non-structural proteins; these markers offered early insights into cross-species transmission (Hu 2010). Building on these foundations, this study took a more integrative approach by applying multiple machine learning algorithms—RF, kNN, and NB—to analyze six key viral proteins (HA, NA, PB1, PB2, PA, NP). The study identified specific physicochemical properties most critical for host specificity across different proteins and key physicochemical attributes for each protein. These included van der Waals volume, polarizability, and charge for HA, and solvent accessibility and electric charge for PB1 and PB2, which are essential for host specificity. By comparing the performance of multiple algorithms, RF emerged as the most effective, achieving superior classification accuracy across all proteins. The study further advanced these findings by employing ranking-based feature selection with RF models to optimize performance. This targeted optimization not only improved computational efficiency but also refined the model's predictive power by isolating the most relevant features for host classification. The multi-algorithm and feature optimization approach offers a robust framework for identifying molecular determinants of host specificity, advancing prior models and contributing significant insights into cross-species transmission dynamics, viral surveillance, and pandemic preparedness.

7.2.5 Applications and Public Health Implications

The insights from this study represent significant contributions to the management of the public health risks posed by influenza and other zoonotic viruses. By identifying host-specific molecular markers, particularly within the HA, NA, and polymerase proteins, this research enhances our capacity to predict cross-species transmission events—a critical factor in mitigating zoonotic spillovers from animal reservoirs to humans. These results could facilitate the establishment of early-warning systems, empowering health authorities to detect emerging viral strains with pandemic potential and initiate proactive responses. Furthermore, the study informs vaccine development strategies by pinpointing key mutations within viral proteins, such as those in the HA RBD. Targeting these mutations could disrupt the virus's ability to bind efficiently to host cell receptors, thereby improving vaccine efficacy. Continuous surveillance of amino acid changes across key proteins would ensure that immunization strategies remain effective against rapidly evolving variants. For example, monitoring antigenic drift in the HA protein would provide valuable guidance for seasonal vaccine updates, reducing the risk of widespread outbreaks. The predictive power of machine learning algorithms enhances public health preparedness, offering real-time data analytics for outbreak prevention and management.

7.2.6 Conclusions and Future Directions

This study underscores the critical role of molecular markers and the potential of machine learning in elucidating the mechanisms of viral host tropism and cross-species adaptation. The HA protein emerged as the most influential determinant for influenza host classification, highlighting its pivotal role in mediating species-specific receptor binding and enabling cross-species transmission. RF algorithms demonstrated robust predictive performance and proved highly effective in analyzing high-dimensional datasets derived from diverse viral proteins. Future research should prioritize experimental validation of the identified molecular markers to confirm their biological significance and clinical applicability. The incorporation of advanced machine learning techniques, such as neural networks, could further enhance prediction accuracy by uncovering complex, non-linear relationships between amino acid features. Expanding the dataset to encompass a wider array of viral strains and host species would improve the model's generalizability, enabling its applicability across different virus families. Adopting this integrated approach would bolster surveillance efforts, enabling the early detection of zoonotic threats and facilitating timely interventions. By advancing our understanding of host-virus interactions and cross-species transmission dynamics, these efforts will enhance global public health preparedness, mitigating the risks associated with future pandemics.

7.3 Case Study 2: Predicting Cross-Species Viral Infection

7.3.1 Case Overview

This study addresses the critical issue of predicting cross-species infection propensities of viruses; this is achieved by focusing on receptor sequence similarities across various host species (Cho and Son 2019). Cross-species infections, particularly zoonotic spillovers, are responsible for numerous global outbreaks, including swine flu, MERS, and Covid-19. This research hypothesizes that evolutionary distances and sequence similarities between receptor proteins act as key determinants in viral transmission across species. By constructing a computational framework based on discriminant analysis, this study aims to systematically predict infection risks, providing a valuable tool for public health surveillance and early-warning systems.

7.3.2 Methodology and Data Analysis

The study collected receptor sequences from 98 host species, comprising 277 amino acid sequences across 18 receptor types from the NCBI database. Each receptor dataset was processed using MSA to ensure accurate sequence comparison and generate predictive variables for modeling. To quantify cross-species infection risk, three metrics were calculated: pairwise evolutionary distances (absolute distance, defined as the number of amino acid substitutions between aligned sequences), relative distances (the ratio of the pairwise distance to the maximum distance within each receptor dataset), and total sequence similarities (the number of matched amino acids in an orthologue sequence relative to the total length of the aligned sequence, including indels).

An initial dataset was then constructed to train, validate, and test a classifier. From this original dataset of 56 sequence pairs, 50 pairs were used for model training and construction through discriminant analysis, while six pairs were reserved for validation. Sequence pairs were classified into two groups, based on susceptibility to viral infection. Group 1 contained high-similarity pairs, indicating increased cross-species infection propensity; Group 2 contained low-similarity pairs, indicating decreased infection propensity. Viruses with unknown or unspecified host receptors were excluded from the dataset, while both zoonotic and non-zoonotic viruses were included. Each receptor dataset included at least one sequence from the natural host (primary reservoir). Group frequencies in the training datasets were 32 for Group 1 and 18 for Group 2; the test datasets contained 4 and 2 sequence pairs for each group, respectively.

A discriminant function model was constructed using Fisher's LDA, with discriminant coefficients calculated to maximize the model's predictive power. The z-score served as a cut-off for classifying host pairs into cross-species infectious and non-infectious groups. Discriminant scores were calculated for all sequence pairs, and group centroids were determined by averaging these scores. Model accuracy was verified through LOOCV (Lachenbruch and Mickey 1968), involving 50 rounds of class predictions with confusion matrices assessed for both original and cross-validated cases. Accuracy was reported as the ratio of correctly predicted sequence pairs to total sequence pairs in the test dataset. Seven rounds of predictions were conducted during testing, and propensity scores were computed to assess infection risk for each sequence pair, providing quantitative measures of cross-species infection propensities.

7.3.3 Key Findings and Interpretation

7.3.3.1 Model Performance and Predictive Accuracy

The study demonstrated a robust computational framework for predicting virus cross-species infection propensities using discriminant analysis. The model achieved 100% classification accuracy in both training and test datasets, as well as in cross-validation, with an AUC value of 1.0, indicating excellent predictive performance. The classification therefore accurately distinguished between infectious and non-infectious host pairs. For example, the zoonotic virus MERS-CoV had a high infection propensity score of 98.759% between camels (*C. dromedaries*) and humans (*H. sapiens*); similarly, SARS-CoV exhibited a high-risk score of 98.495% between cats (*F. catus*) and humans (*H. sapiens*). The model identified the Nipah virus as having a high infection propensity between swine (*S. scrofa*) and humans (*H. sapiens*), with a score of 88.572%. The model's ability to reflect real-world interspecies infection patterns highlights its reliability and practical application in zoonotic risk assessments.

7.3.3.2 Role of Sequence Similarity and Evolutionary Distance of Receptors in Host Susceptibility

The study found that sequence similarity between host receptor proteins was the most influential predictor of cross-species infection. Host pairs with higher sequence similarity showed a greater likelihood of viral transmission. Evolutionary distance also played a significant role, with more closely related species—based on receptor sequence similarity—showing higher infection risk. For instance, high similarity between the receptors used by avian influenza virus infecting domestic poultry and wild birds resulted in a high propensity score, highlighting the ease of viral transmission within avian species. This example underscores the importance of receptor compatibility in determining host switching. The Nipah virus demonstrated a strong propensity for transmission between swine and humans (*S. scrofa-H. sapiens*) as well as between bats and humans *(M. brandtii-H. sapiens)*, confirming the predictive power of receptor similarity in zoonotic risk.

Taken together, variables derived from sequence similarity and evolutionary distance act as either positive or negative influences on species-specific susceptibility. In this study, data with high sequence similarity and close evolutionary distance produced a high infection propensity. Variation among receptor sequences may impact host susceptibility, but exceptional patterns exist. For example, some cases do not follow linear species distances due to nuanced differences in host specificity. Although the model achieved strong cross-validation results (83.3% accuracy for the test dataset), the incomplete coverage of receptor sequence polymorphisms suggests further improvements are needed. Thus, this model should be cautiously applied to cases where more subtle host specificities exist. The assumption should be made that as evolutionary distance increases and sequence similarity decreases, the probability of sharing susceptibility-associated sequence properties also decreases. The discriminant coefficient (7.026) of similarity within sequence pairs—which positively influenced host susceptibility—confirmed the importance of sequence polymorphisms in improving prediction accuracy. These findings indicate that the model could be enhanced by incorporating more receptor sequence data.

7.3.3.3 A Computational Framework and Web-Based System for Predicting Virus Cross-Species Infection Propensity

The computational framework developed in this study consists of four stages: sequence analysis, classification model construction, model-based prediction, and calculation of infection propensity scores. The framework utilized sequence similarity and evolutionary distances between host receptor proteins as primary variables for predicting cross-species infection. Discriminant analysis classified each host pair into either an infectious or non-infectious group, and propensity scores quantified the infection risk for each pair. This model successfully identified key variables predicting viral transmission and represents a practical tool for assessing infection potential between different host species. The high performance of this framework demonstrates its potential for enhancing zoonotic infection risk prediction and supporting public health interventions aimed at preventing viral spillovers from animal reservoirs to human populations.

Additionally, this study developed ViCIPR (Virus Cross-species Infection Propensity Resource), a server-based tool that enables users to assess the cross-species infection propensities of specific viruses. ViCIPR allows users to search for sequence homology and calculate infection propensity (Infectindex) based on query sequences, which can be submitted by direct entry or through FASTA file upload. Using its "Sequence Similarity Scoring System," ViCIPR compares the query sequence against three reference databases—nucleotide genomes, CDS, and proteins. It then displays selectable secondary (recipient) host species based on the sequence similarity results and presents the calculated Infectindex for the input query in real time. This ultimately allows for efficient evaluation of cross-species infection risk.

7.3.4 Comparative Insights and Innovations in Zoonotic Risk Prediction

This case study adopted a comprehensive approach by incorporating a diverse range of receptor types and comparing them across multiple host species, facilitating predictions that encompass a wider variety of zoonotic viruses. The discriminant analysis model developed in this study integrates evolutionary distance and sequence similarity, enabling it to capture both host-virus compatibility and zoonotic risk potential—dimensions that were less comprehensively explored in prior models. Furthermore, this study introduces ViCIPR, a server-based tool designed to assess infection propensity by comparing query sequences against reference nucleotide and protein sequence databases. ViCIPR offers real-time scoring of infection propensity, enhancing accessibility and enabling practical, rapid assessment. This innovation provides a valuable resource for viral surveillance and public health monitoring. By offering a user-friendly and integrated platform, this study not only builds upon earlier findings but also extends them, supporting timely zoonotic risk assessment and strengthening preparedness for cross-species transmission events.

7.3.5 Applications and Public Health Implications

The predictive framework developed in this study provides critical insights for managing zoonotic threats and enhancing pandemic preparedness. The identification of host-specific molecular markers and calculation of infection propensity scores based on receptor sequence similarity enables public health authorities to forecast and monitor potential cross-species transmission events. This approach supports targeted surveillance efforts, facilitating the early detection of viruses with pandemic potential, such as influenza viruses, SARS-CoV and Nipah virus. Furthermore, this framework can inform vaccine development by identifying receptor-binding sites critical for viral adaptation, thereby guiding immunization strategies. Continuous updates to receptor sequence databases will enhance the model's capacity for the real-time monitoring of viral evolution, ensuring that interventions remain effective amidst dynamic viral landscapes. With its scalability and adaptability, this predictive framework is an invaluable resource for comprehensive viral surveillance across diverse ecological settings, reinforcing global health initiatives to pre-empt zoonotic threats.

7.3.6 Conclusions and Future Directions

This study underscores the importance of receptor sequence analysis and discriminant modeling in predicting cross-species viral infections. While the model demonstrated exceptional accuracy and practical utility, it also requires further refinement. Expanding receptor sequence datasets and incorporating structural data, such as tertiary protein configurations, would provide a more nuanced understanding of virus–host interactions. Future research should integrate additional factors, including immune responses, viral transmission pathways, and environmental variables, to enhance the model's predictive precision. Further insights into viral dynamics will arise from the growing availability of receptor data, experimental validation of estimated cross-species infection propensities, advances in computational capabilities, and the adoption of complex machine learning models, such as neural networks. This framework represents a robust foundation for developing advanced predictive tools to monitor viral evolution and cross-species transmission, ultimately fortifying global health preparedness against emerging zoonotic diseases.

References

Allen JE, Gardner SN, Vitalis EA, Slezak TR. Conserved amino acid markers from past influenza pandemic strains. BMC Microbiol. 2009;9:77.

Bateman AC, Busch MG, Karasin AI, Bovin N, Olsen CW. Amino acid 226 in the hemagglutinin of H4N6 influenza virus determines binding affinity for α2, 6-linked sialic acid and infectivity levels in primary swine and human respiratory epithelial cells. J Virol. 2008;82:8204–9.

Bayes T. An essay towards solving a problem in the doctrine of chances. Philos Trans R Soc Lond. 1763;53:370–418.

Breiman L. Random forests. Mach Learn. 2001;45:5–32.

Bussey KA, Desmet EA, Mattiacio JL, Hamilton A, Bradel-Tretheway B, Bussey HE, Kim B, Dewhurst S, Takimoto T. PA residues in the 2009 H1N1 pandemic influenza virus enhance avian influenza virus polymerase activity in mammalian cells. J Virol. 2011;85:7020–8.

Chen Z, Zhao P, Li F, Leier A, Marquez-Lago TT, Wang Y, Webb GI, Smith AI, Daly RJ, Chou K-C, Song J. iFeature: a python package and web server for features extraction and selection from protein and peptide sequences. Bioinf. 2018;34:2499–502.

Cho M, Son HS. Prediction of cross-species infection propensities of viruses with receptor similarity. Infect Genet Evol. 2019;73:71–80.

Cohen M, Zhang XQ, Senaati HP, Chen HW, Varki NM, Schooley RT, Gagneux P. Influenza a penetrates host mucus by cleaving sialic acids with neuraminidase. Virol J. 2013;10:321.

Dubchak I, Muchnik I, Holbrook SR, Kim SH. Prediction of protein folding class using global description of amino acid sequence. Proc Natl Acad Sci. 1995;92:8700–4.

Dubchak I, Muchnik IB, Kim SH. Protein folding class predictor for SCOP: approach based on global descriptors. ISMB. 1997:104–7.

Fisher RA. The use of multiple measurements in taxonomic problems. Ann Eugenics. 1936;7:179–88.

Fisher RA. Statistical methods for research workers. In: Breakthroughs in statistics: methodology and distribution. Springer; 1970. p. 66–70.

Fix E, Hodges Jr. JL. Discriminatory Analysis; Nonparametric Discrimination: Consistency Properties USAF School of Aviation Medicine, Randolph Field, TX, Project Number: 21-49-004, Report Number: 4; 1951.

Hall MA. Correlation-based feature selection for machine learning (Doctoral dissertation, The University of Waikato); 1999.

Hastie T, Tibshirani R, Friedman J. The elements of statistical learning: data mining, inference, and prediction. 2nd ed. Springer; 2009.

Hu W. Novel host markers in the 2009 pandemic H1N1 influenza a virus. J Biomed Sci Eng. 2010;3:584–601.

Hueffer K, Parker JS, Weichert WS, Geisel RE, Sgro JY, Parrish CR. The natural host range shift and subsequent evolution of canine parvovirus resulted from virus-specific binding to the canine transferrin receptor. J Virol. 2003;77:1718–26.

Imai M, Kawaoka Y. The role of receptor binding specificity in interspecies transmission of influenza viruses. Curr Opin Virol. 2012;2:160–7.

Kesimer M, Scull M, Brighton B, DeMaria G, Burns K, O'Neal W, Pickles RJ, Sheehan JK. Characterization of exosome-like vesicles released from human tracheobronchial ciliated epithelium: a possible role in innate defense. FASEB J. 2009;23:1858–68.

Kwon E, Cho M, Kim H, Son HS. A study on host tropism determinants of influenza virus using machine learning. Curr Bioinforma. 2020;15:121–34.

Lachenbruch PA, Mickey MR. Estimation of error rates in discriminant analysis. Technometrics. 1968;10:1–11.

Li ZR, Lin HH, Han LY, Jiang L, Chen X, Chen YZ. PROFEAT: a web server for computing structural and physicochemical features of proteins and peptides from amino acid sequence. Nucleic Acids Res. 2006;34:w32–7.

Matthews BW. Comparison of the predicted and observed secondary structure of T4 phage lysozyme. Biochim Biophys Acta. 1975;405:442–51.

McAuley JL, Corcilius L, Tan HX, Payne RJ, McGuckin MA, Brown LE. The cell surface mucin MUC1 limits the severity of influenza a virus infection. Mucosal Immunol. 2017;10:1581–93.

McAuley JL, Gilbertson BP, Trifkovic S, Brown LE, McKimm-Breschkin JL. Influenza virus neuraminidase structure and functions. Front Microbiol. 2019;10:39.

Mehle A, Dugan VG, Taubenberger JK, Doudna JA. Reassortment and mutation of the avian influenza virus polymerase PA subunit overcome species barriers. J Virol. 2012;86:1750–7.

Roffo G, Melzi S. Ranking to learn: Feature ranking and selection via eigenvector centrality. In: New Frontiers in Mining Complex Patterns: 5th International Workshop, NFMCP 2016, Held in Conjunction with ECML-PKDD 2016, Riva del Garda, Italy, September 19, 2016, Revised Selected Papers 5, 19–35. Springer International Publishing; 2017. arXiv:1704.05409v1 [cs.CV].

Schneider-Schaulies J. Cellular receptors for viruses: links to tropism and pathogenesis. J Gen Virol. 2000;81:1413–29.

Tarendeau F, Crepin T, Guilligay D, Ruigrok RW, Cusack S, Hart DJ. Host determinant residue lysine 627 lies on the surface of a discrete, folded domain of influenza virus polymerase PB2 subunit. PLoS Pathog. 2008;4:e1000136.

Tukey JW. Comparing individual means in the analysis of variance. Biometrics. 1949;5:99–114.

Vines A, Wells K, Matrosovich M, Castrucci MR, Ito T, Kawaoka Y. The role of influenza a virus hemagglutinin residues 226 and 228 in receptor specificity and host range restriction. J Virol. 1998;72:7626–31.

Wan H, Perez DR. Amino acid 226 in the hemagglutinin of H9N2 influenza viruses determines cell tropism and replication in human airway epithelial cells. J Virol. 2007;81:5181–91.

Xu D, Newhouse EI, Amaro RE, Pao HC, Cheng LS, Markwick PRL, McCammon JA, Li WW, Arzberger PW. Distinct glycan topology for avian and human sialo-pentasaccharide receptor analogues upon binding different hemagglutinins: a molecular dynamics perspective. J Mol Biol. 2009;387:465–91.

Part IV

Summary and Additional Resources

Summary 8

This book was written to introduce the field of viroinformatics. Viroinformatics emerged from the intersection of the long-established discipline of virology and the relatively new field of bioinformatics, all within the broader context of public health. Therefore, it is self-evident that our ultimate goal aligns with the fundamental purpose of public health: to protect and improve human health. However, beyond this grand, almost philosophical necessity, it is essential to incorporate practical aspects; public health, by its very nature, must be practical. While this book provides detailed explanations throughout, this summary aims to capture the key points and overarching themes covered in the text.

Chapter 1 begins with a general yet essential overview of viruses, offering a broad understanding of these complex entities. The field of virology has a long and rich history, with scientists having studied viruses for centuries. However, compared to the existence of viruses themselves, virology's history is relatively brief. This gap highlights the fact that our understanding of viruses is still in its infancy, suggesting that extensive and profound discoveries relating to viruses will continue for many years to come. In the grand context of virology, the vast body of accumulated knowledge represents only a small portion of what there is to know. This chapter serves as a starting point, offering a foundational overview, yet it only scratches the surface of what remains to be explored. Towards the end of Chap. 1, the book addresses the coronavirus, which recently became a global public health crisis. This section emphasizes the urgency of the current situation, providing readers with an indirect sense of the global nature and ongoing challenges of viral outbreaks. While general knowledge serves as a solid foundation, it is clear that this introductory overview is only a stepping stone for more specialized investigations. The book emphasizes that while the general information provided in Chap. 1 is an essential foundation, it serves as an entry point rather than a comprehensive analysis of viruses. More advanced and specialized research is necessary to fully understand the complexities of viral biology, evolution, and pathology. This chapter lays the groundwork for deeper exploration, setting the stage for the sophisticated methodologies and cutting-edge discoveries that will follow in subsequent chapters.

Chapter 2 introduces bioinformatics, a dynamic and rapidly evolving field that has recently undergone tremendous growth, particularly since the completion of the Human Genome Project. In the span of just a few decades, bioinformatics has transformed from a niche discipline into a cornerstone of modern biological research. One of the most remarkable developments in this field has been the integration of big data and AI, both of which have broadened the

H. S. Son, M. Cho, *Viroinformatics and Computational Virology in Public Health*,
https://doi.org/10.1007/978-981-96-5757-5_8

scope and impact of bioinformatics across a wide array of scientific disciplines. The sheer volume of biological data now being generated—thanks to advances in sequencing technologies like Next Generation Sequencing—has necessitated the development of more sophisticated tools and computational methods for data analysis. Furthermore, these advances have also contributed to a shift in the terminology and frameworks used within bioinformatics. As hardware has improved and the capacity for large-scale data production has grown exponentially, methods for processing, analyzing, and interpreting biological data have evolved in tandem. Given the rapid pace of technological change, it is worth noting that any specific tools or methodologies described today may quickly become outdated, as the field advances. This constant evolution raises the question of whether there is much value in detailing technologies that could become quickly obsolete. Despite this, the importance of understanding the current landscape cannot be overstated. Therefore, Chap. 2 focuses on providing a basic yet essential introduction to bioinformatics, grounding the reader in the fundamental principles and techniques that form the foundation of this transformative field.

Chapter 3 focuses on phylogenetics, a critical discipline in biological research that plays a central role in understanding viruses and their evolution. While the debate over whether viruses should be considered "living organisms" persists, this question becomes less relevant in the context of evolution. Evolution is a universal process that applies to all biological entities, regardless of whether they are classified as living or not. In the case of viruses, although they lack certain characteristics of life, such as a cellular structure, they still undergo evolutionary changes through mechanisms of mutation and adaptation. These processes, although sometimes described simply as "sudden changes," follow a systematic, gradual path driven by evolutionary pressures. Phylogenetics is therefore an indispensable tool in viral research, as it allows scientists to trace evolutionary relationships, track virus transmission patterns, and understand the molecular underpinnings of viral evolution. This chapter concludes with a discussion of the evolutionary relationship between lentiviruses in humans and non-human primates, highlighting how phylogenetic analysis can shed light on cross-species infections, a critical area of study for understanding the origins and spread of emerging infectious diseases.

Chapter 4 provides an in-depth exploration of various indices and concepts of codon usage analysis, a key aspect of understanding the molecular dynamics of gene expression. This chapter introduces several methods that allow researchers to evaluate codon usage preferences, diversity, and efficiency within genes. One of the primary tools discussed is "Base Composition," which analyses the relative proportions of the four DNA bases (A, T, C, G), providing insights into the overall structure of the genome. "RSCU" (Relative Synonymous Codon Usage) is another critical index that measures how often synonymous codons are used within a gene, which can affect the efficiency of protein synthesis. "Fop" (Frequency of Optimal Codons) is employed to assess the presence of preferred codons that improve translation speed and accuracy. The "CBI" (Codon Bias Index) is used to evaluate the degree of codon preference within a gene, while the "CAI" (Codon Adaptation Index) compares the patterns of codon usage between genes and the host genome, providing insights into gene optimization. "ENC" (Effective Number of Codons) analysis quantifies codon usage diversity, while the "ICDI" (Index of Codon Diversity) offers a method for assessing natural variability in codon selection. Finally, "ER" (Evolutionary Rate) quantifies the evolutionary dynamics shaping codon usage, reflecting both genetic divergence over time and the selective pressures acting on genes. These indices are essential for understanding how codon usage influences gene translation efficiency and evolution over time. This chapter further examines the evolution of codon usage in viruses, including a case study on Influenza A, demonstrating how these tools contribute to virus research. In conclusion, the chapter emphasizes the importance of codon usage analysis in both gene and virus research and suggests potential future research directions.

Chapter 5 explores AI algorithms and their applications. This chapter introduces a comprehensive range of AI algorithms, explaining their characteristics and applicability. Key algorithms discussed include regression analysis, SVMs, kNN, clustering techniques, decision trees, RFs, neural network-based algorithms (e.g., CNN, RNN, LSTM), GANs, and transformer models. These algorithms play a crucial role in data analysis, prediction, classification, and pattern recognition and have significant applications in virology research. AI is widely used in virus classification, antiviral and vaccine development, diagnostic improvements, predictive modeling, epidemiological analysis, and personalized treatment strategies. This chapter provides examples of how AI has been applied in research, particularly how machine learning helps in understanding host-specific virus infection characteristics, and underscores AI's growing importance in virology. The chapter concludes by discussing the future potential of AI advances and their future applications in virology.

Chapter 6 builds upon the analytical methods and algorithms discussed in earlier chapters to explore how these tools can be applied to study viral evolution. This chapter delves into the mechanisms and forces that drive viral evolution, offering insights into how viruses adapt, persist, and sometimes emerge as new threats to public health. The chapter highlights several analytical methods, including phylogenetic analysis (introduced in Chap. 3) and codon usage analysis (covered in Chap. 4), providing readers with a comprehensive toolkit. Examples of the use of these tools are provided to demonstrate how viral genomes evolve and adapt. Among the methods discussed are phylogenetic analysis, which traces the evolutionary history and relationships between viruses, and nucleotide composition analysis, which examines the specific makeup of viral genomes to uncover patterns of mutation and adaptation. The chapter also focuses on codon usage pattern analysis, a crucial method for understanding how the selection of specific codons can influence viral evolution. By evaluating factors such as codon frequency, selective pressure, and mutation pressure, researchers can track viral evolutionary trends and gain insights into how viruses adapt to different environments. To illustrate the application of these tools, the chapter presents case studies on HPV oncogenes and polyomavirus LT-Ag genes, each analyzing codon usage patterns and evolutionary dynamics across various genotypes or host species. These case studies provide a deeper understanding of the complex processes driving viral evolution, with important implications for public health. In conclusion, the chapter underscores the significance of codon usage analysis in studying viral evolution and its application in future research.

Chapter 7 focuses on the application of analytical methods and algorithms introduced in earlier chapters to predict viral host specificity and cross-species infections. It highlights a range of analytical methods designed to predict how viruses interact with specific hosts and assess the likelihood of transmission between species. These techniques, including feature extraction, machine learning analysis, performance evaluation, and optimization, offer powerful tools for understanding complex virus-host dynamics. Feature extraction identifies key characteristics of viruses and hosts that influence interactions, while machine learning algorithms, such as SVMs and RFs, can process vast amounts of data to predict potential host ranges and the risk of cross-species transmission. The chapter presents real-world case studies on influenza virus host specificity and the prediction of interspecies infection of zoonotic viruses to demonstrate how these methods can be applied in practice. By utilizing data from previous outbreaks, these studies highlight how machine learning models can accurately predict which animal species or humans are most likely to be affected by new strains of viruses, offering valuable insights for early warning systems. The chapter also explores the public health implications of these predictive methods, emphasizing their potential to inform preventative measures and improve response strategies during viral outbreaks. Additionally, the chapter outlines future directions for research in viral host specificity and transmission prediction, encouraging continued innovation in this important area. The author's own research informs much of the book, but the examples are intended to inspire, not impose rigid methodologies. Scientific research, as always, should be flexible, open to refinement, and driven by the goal of continuous improvement.

We are still navigating the complex and evolving interactions between virology, AI, and evolutionary biology. This area of study is not merely driven by scientific curiosity but is vital for addressing ongoing and future challenges to public health. As viruses continue to evolve and adapt, they present persistent threats that require us to better understand their behavior and develop more effective strategies for prevention, treatment, and control. The integration of virology and modern technologies, such as AI, offers new opportunities to predict and respond to viral threats in ways that were previously unimaginable. These advances provide the tools to uncover patterns in viral evolution, which could help anticipate the emergence of new viral strains and improve public health responses. It is essential that the field continues to advance to keep pace with viral evolution and emerging threats, ensuring that we are prepared for the challenges of tomorrow.

The research and analysis presented in this book aim to contribute to the future of virology and viroinformatics and their roles in public health. By exploring viral behavior, evolution, and interactions with hosts, this work seeks to expand our understanding of these processes and support the development of more effective solutions for combating viral diseases. As our understanding of viruses and their impact on human society continues to grow, the tools and methods we use will evolve to address emerging challenges. This book has been written with the hope of contributing to such efforts.

Index

H. S. Son, M. Cho, *Viroinformatics and Computational Virology in Public Health*,
https://doi.org/10.1007/978-981-96-5757-5

MIX
Papier aus verantwortungsvollen Quellen
Paper from responsible sources
FSC® C105338

If you have any concerns about our products,
you can contact us on
ProductSafety@springernature.com

In case Publisher is established outside the EU,
the EU authorized representative is:
Springer Nature Customer Service Center GmbH
Europaplatz 3, 69115 Heidelberg, Germany

Printed by Libri Plureos GmbH
in Hamburg, Germany